MW00650669

OPERATIONAL EXCELLENCE

OPERATIONAL EXCELLENCE

Journey to Creating Sustainable Value

JOHN S. MITCHELL

WILEY

Published by John Wiley & Sons, Inc., Hoboken, New Jersey
Published simultaneously in Canada

Library of Congress Cataloging-in-Publication Data:

Mitchell, John S., 1937-
Operational excellence : journey to creating sustainable value / by John S. Mitchell.
pages cm
Includes index.
Summary: "Provides Operational Excellence strategies and implementation details for enterprises at all levels of maturity, from those with programs in place to those looking to improve performance"– Provided by publisher.
ISBN 978-1-118-61801-1 (hardback)
1. Production management. 2. Industrial management. 3. Quality control. 4. Total quality management. I. Title.
TS155.M6165 2015
658.5–dc23

2014025096

Printed in the United States of America

10 9 8 7 6 5 4 3 2

CONTENTS

BIOGRAPHY OF JOHN S. MITCHELL

During a professional career of over 40 years, John Mitchell has held a wide range of executive, managerial, and technical positions in industry. He has been a strong and visible advocate for the development and implementation of business, technical, and operating strategies for Operational Excellence, Physical Asset Management, and Reliability and Maintenance (R&M) excellence. During his career, he has delivered numerous presentations and workshops throughout the world, stressing the necessity and financial and business benefits of optimizing work processes and technology in his areas of expertise. In semiretirement, he provides mentoring and coaching for organizations committed to improving operational productivity and effectiveness.

Mitchell has authored over 100 technical papers and articles, detailing the technology, financial, and business benefits of Operational Excellence, Physical Asset Optimization, Profit-Centered Maintenance, reliability improvement, condition monitoring, and assessment. Mitchell has authored the books "*The Physical Asset Management Handbook,*" currently in its fourth edition, and "*An Introduction to Machinery Monitoring and Analysis.*"

He is a graduate of the U.S. Naval Academy, Annapolis Maryland.

PREFACE
JOURNEY TO OPERATIONAL EXCELLENCE

This book is addressed to three constituencies:

Executives and senior management: the text provides a solid definition of the concept and principles of Operational Excellence, the business/mission benefits and the value it will produce, and the importance leading enterprises have placed on its successful implementation. This information is contained in the first five chapters along with principles, requirements, basics of implementation, and commitment necessary for success. Consider the five chapters as an overly long executive summary of the details that follow.

Leaders and third parties chartered to lead, manage, and facilitate an Operational Excellence initiative: specific information detailing the establishment and effective implementation of the program. An Operational Excellence program produces results meeting all expectations and, of course, makes them look good in the process.

People within an enterprise participating in Operational Excellence: this group will find definitive principles and a detailed implementing roadmap to assure all are on the same page, working effectively to common objectives. Expectations, requirements, and mutual responsibilities are fully identified and aligned to assure maximum success.

EVOLUTION OF OPERATIONAL EXCELLENCE

The concepts and details that follow have been developed and refined over more than a decade of workshops and interactions with enterprises and people directly involved in seeking maximum effectiveness and return from their efforts. This text will provide guidance and great value for executives, management, leaders, and implementers of Operational Excellence.

The path to this book began in the early 1990s. The author and a good friend, both with land-based production operations and marine experience, were attempting to discern why industrial plant maintenance, rather than the essential, profit-making part of the business that it could and must be, was typically considered a necessary evil and cost to be controlled. In many operating enterprises, maintenance was primarily directed to restoring failures reactively rather than proactive avoidance, elimination, and improvement. Furthermore, many advances in technology and practice proven to improve performance and value gained from physical production assets were not viewed in the same light as investments for improving energy and process efficiency. While the latter was presumed to generate a business return, the former was most often thought of as rather costly luxuries to be considered only when times were good. Equally important, why did performance-limiting friction exist at the working level between the operations and maintenance functions in most industrial operating enterprises?

In many operating enterprises, operations typically felt that production/mission compliance was first, foremost, and above all. Operations typically considered maintenance a service supplier: someone to call when bad things happened. This perspective and its corollary "keep the customer happy" were shared by many in maintenance. Although typical within many operating enterprises, the "keep the customer happy" premise is particularly unproductive when considering what obligations a customer, operations in this case, has to a supplier. What obligations do you have to the supplier of the tires on your automobile? Operations would say that maintenance was often unrealistic, requesting a premature halt to production to correct what operations considered minor problems or to conduct preventive tasks operations considered of secondary importance to production output. Maintenance typically replied that production too often operated equipment carelessly, didn't really care for or about equipment until something went wrong, and then applied extreme pressure to restore operation as quickly as possible.

In the marine industry, operators and maintainers are one and the same—these tensions didn't exist. Even if it did, open ocean swimming in creature-filled waters isn't an attractive alternate in the event of a major failure!

A concept called *profit-centered maintenance* was the initial outcome of the discussions. Profit-centered maintenance advocated adopting a value, investment, return, and continuous improvement mentality, rather than the less-effective cost control to budget where there are actually disincentives for improvement.

A leading operating company had adopted a team-based profit-centered mentality within a budgetary system. It appeared to work very well, with numerous operating and business benefits. Many who reviewed profit-centered maintenance thought the concept as a good idea; some felt it didn't go far enough. One person stated that maintenance in his company was considered, and always would be considered, a service and cost—the profit-centered philosophy would never be accepted at any level of the company, certainly not by executive and financial management.

By the late 1990s, the profit-centered idea had advanced into the initial concept of physical asset management. In addition to the essentials of profit-centered maintenance, physical asset management expanded to emphasize the cooperative organization, improvement-oriented work culture, human performance excellence, reliability, and risk imperatives. All are necessary within a larger, multifunction, operating process to gain optimum sustainable performance and greatest business results.

A book, written by the author and first published in early 2000, detailed the value-based principles of physical asset management. The book made the case that the key challenge facing operating enterprises wasn't technology or adoption of optimizing practices—all had been proven beyond doubt—but rather the business value that made them essential for success. This was the fundamental premise of physical asset management.

During the first decade of the twenty-first century, physical asset management, under several naming variants, gradually devolved into a more sophisticated name for maintenance. The profit center value concept, operations maintenance cooperation, empowered workforce, human performance excellence, and a number of other elements considered essential for optimum business/mission performance and effectiveness were subsumed by heavy concentration on administrative controls, maintenance management, and practice.

Late in the decade, the term *Operational Excellence* began to appear, primarily from leading operating enterprises as their guiding concept for attaining maximum effectiveness, overall operations, and business and mission success. Examining the principles of Operational Excellence as expressed by industry leaders, a number of essentials codified earlier concepts. Engaged visionary leaders, value focus, a working culture committed to excellence in all activities and continuous improvement, optimizing risk and reliability as business imperatives, cooperative, committed ownership throughout the organization, and empowered multifunction improvement action teams to name a few.

Viewing the principles, it became clear that Operational Excellence could be considered the latest stage in an evolution that began almost 20 years earlier with the concept of profit-centered maintenance. In addition to the profit and value focus, Operational Excellence includes production and business/mission optimization for greatest effectiveness and success. It is fully inclusive, compatible with, and supportive of asset management. In fact, the organizational, administrative, and control system specified in the ISO55000 series, Asset Management, are totally applicable to and should be incorporated within Operational Excellence. Think of Operational Excellence as a major river to business/mission success. Lean Six Sigma, energy and control performance excellence, asset management and asset performance excellence, profit-centered maintenance, and other broadly accepted functional improvement practices, production, and business system optimization are tributaries! All are included in, contributing to, and driving Operational Excellence to gain greatest value for the enterprise.

Physical asset management workshops, delivered throughout the world by the author beginning in the late 1990s and continuing during the first decade of the 2000s, gradually shifted emphasis to Operational Excellence. The book mentioned previously, likewise evolved through four editions, increasing emphasis on the value and human elements that are essential for a successful operating enterprise.

GROWING AWARENESS OF OPERATIONAL EXCELLENCE

As time progressed, several facts became apparent. There is growing awareness of Operational Excellence at executive levels of operating enterprises. Many operating enterprises are implementing or considering Operational Excellence. Most of these have a dedicated executive at the vice president or director level overseeing the effort. Some may call the concept by a different name—the powerful principles remain the same. With this stated, definitions are evolving, and there are few generally accepted implementing details to provide guidance for those assigned responsibility. In one major corporation, overall guidelines for Operational Excellence signed by the CEO were so general and with a number of inconsistencies that individual business units within the corporation developed differing implementations. Books and articles on Operational Excellence have been focused into areas such as the organization, process flow, and alignment of management systems; few address the totality. Furthermore, there is a major disagreement on overall strategy. Should the operating strategy of the enterprise be directed to business growth, improving value gain, continuous improvement, or some combination? All these issues will be addressed and answered.

Many participants at Operational Excellence workshops stated their company had initiated an Operational Excellence program or was considering a program without any details to define the beast beyond the compelling term! The great majority were present to learn and assure the program they were developing, or might be called on to develop, would meet expectations quickly and effectively, thereby providing executive management confidence in Operational Excellence and their personal value, not necessarily in that order! A few stated they had Operational Excellence initiatives in place and were looking for ideas that would make their efforts more effective and successful.

By now, many are wondering where safety is in this concept. In fact, safety and Operational Excellence are interlocking, identical programs. The same procedural basis, working culture, organizational and individual commitment, learning, and continuous improvement required for safety must exist within Operational Excellence. Much more about this will be revealed as the story unfolds. There are many supporting functions. Several have been named, and more will be revealed in the text. Human relations (HR) and personnel and change (improvement) management are highly important. It is the same for engineering, finance, and information technology (IT). For all, the text will focus on requirements and function within Operational Excellence. Several specific practices such as Lean Six Sigma will be identified and their application discussed; many details are left to subject-specific texts.

There you have it, a brief explanation of where we will be going and a bit about the origin. Hope the text will be informative and create value for you.

ACKNOWLEDGMENTS

Developing the concepts and ideas for a book of this type begins from experience, identifying and solving problems, discussions, comments, and suggestions with many, many people. To all I have come in contact with over the years, participants in projects, plant assessments, technical conferences, physical asset management, and operational excellence workshops—my great thanks for sharing your knowledge, your questions, comments, and insight toward building this body of knowledge.

My great appreciation to BJ Lowe for taking on the Handbook of Physical Asset Management more than ten years ago—supporting its distribution and improvement through four editions over the intervening years.

Thanks to Terry O'Hanlon and reliabilityweb.com for sponsoring the initial reliability, and Physical Asset Management workshops, and later the Operational Excellence workshops from which a great deal of this material was initially presented and refined.

Thanks to Dr. Peter Martin for continuing encouragement, several thought provoking, save the world conversations over dinner that generated ideas, and your excellent contribution to this book.

My appreciation to Paul Barringer for your friendship, many interesting e-mails, your continuing great contribution to the reliability field, and excellent and thought provoking contribution to this book.

Thanks to Grahame Fogel for many challenging discussions over the years that led directly to strengthening the concepts presented in this book; making certain they would add to and reinforce programs and practices that have proven effective. Thanks to Marty Moran, your great contribution during many days spent jointly developing the assessment template that provides specific guidelines for the necessities of Operational Excellence.

My appreciation to Dr. P.J. Vlok for allowing me to test many ideas during your reliability, and asset management conferences. Who could have ever guessed that an impromptu trip to the Hollywood sign would have led to a ten year plus friendship! To you and Grahame, many thanks for your great hospitality during memorable visits to South Africa.

Great appreciation to my very old friend Willie Gerritts for support and encouragement over the years. To Willie, and his colleague Rohann Botha, thanks to you both for providing the encouragement to commit thoughts to print, and then taking time to read, and comment on the manuscript as it developed. Congratulations to Willie for your fine work in Operational Excellence. SASOL is fortunate to have highly motivated and committed people like you contributing mightily in their journey to greatness.

Many thanks and great appreciation to Heinz Bloch whom I have been privileged to know and consider a great friend for more years than we both care to count. Special thanks for making the crucial connection with John Wiley that resulted in this book.

Thanks to Jack Kulp, a good friend, founder of Traffix Devices and a highly successful businessman, for a conversation that led to crucial concepts presented in Chapter 2.

Great appreciation to my brother, Bill Mitchell for your review, and comments from the executive suite you occupied most successfully for many years. Your entire extended family is very proud of you.

Thanks to Bob Esposito and Michael Leventhal at John Wiley for accepting the proposal for this book. Special thanks to Michael for your consideration and answering questions as the manuscript developed. Thanks also to the reviewers who concluded that the proposed book would be a worthwhile endeavor for John Wiley; trust the result meets your expectations.

When a manuscript is finally complete, the editorial process is all important to elevate an engineer's thoughts, and often fractured grammar, to a polished final form. Thanks to Jayshree Saishankar at Wiley, for her involvement and support with this book. Very special thanks to Dhanalakshmi Ram, the copyeditor for the book, Aishwarya Daksinamoorty and Kartika Rajendran, Project Managers and her colleagues in SPi Global who helped craft the draft manuscript into final pages of print which you see within these covers.

Finally, and certainly not least, thanks to my wonderful, and very patient wife Pat for whom I will have to think of another answer to the question: "How much longer are you going to sit behind that computer?"

JOHN MITCHELL

September 2014

INTRODUCTION

Peter G. Martin
Vice President Invensys

Operational Excellence is certainly not a new concept. Although perhaps not using the specific term, operating enterprises have been striving to implement Operational Excellence across their production and manufacturing sites for decades. Operational Excellence has meant that an enterprise is running their operations in the best possible manner. Of course, the question is "what does best mean?" Up until a just few years ago, most executives striving for Operational Excellence were working to have the most efficient operations. Therefore, "best" in this context referred to superior operational efficiency. Typically, high efficiency was measured in terms of the following:

- Actual production throughput to maximum potential production throughput
- Actual energy consumption to minimum potential energy consumption
- Actual material consumption to minimum potential material consumption
- Minimum safe headcount.

All of the preceding were to be achieved while maximizing safety and minimizing environmental impact. Therefore, the primary objective of operating enterprises has been to maximize production throughput safely while minimizing energy and material consumptions and human costs. Attaining Operational Excellence has been and continues to be an ongoing challenge for all industrial and mission-centered operating enterprises. In many cases, it is a matter of survival.

Operational Excellence: Journey to Creating Sustainable Value, First Edition. John S. Mitchell.

Over the past decade, there has been a subtle but highly impactful shift in terms of what industrial leaders are recognizing they must have in order to attain real, sustaining Operational Excellence. The traditional concept focused on increasing operational efficiency has proven to be inadequate. Maximizing efficiency, as daunting a challenge as that is, is no longer enough. This may appear like a fairly limited shift in perspective, but it is not. Traditionally, business executives focused on profitability, and operating personnel focused on efficiency. There was a clear separation between business and operations, and business fared quite well. But over the past decade, this separation of responsibility has led to underperformance. It is important to understand why this change has occurred and what the impact of this change is in terms of how to achieve business/mission Operational Excellence by this new definition.

I was involved with a project spanning much of the past two decades to try to determine what business executives are looking for out of their operations talent and technology. During this period, over 2000 executives from multiple production and manufacturing enterprises were involved in structured interviews, focus groups, and other information-gathering sessions. The objective of this project was to help determine how operations and technologies might be more effectively used to meet executives' desires and expectations. It is certainly beyond the scope of this book to go into all aspects of what these executives communicated, but one aspect of this project is particularly revealing when discussing Operational Excellence. That is, most executives involved in the project conveyed that one of their biggest concerns was that they had no real-time visibility into where and when they were making and losing money across their enterprises and in their operations. Many expressed frustration that they typically did not know if they were having a good or bad month until 5 or 6 days after the end of the month when they received monthly closing reports from the enterprise resource planning (ERP) system. They indicated that they were often surprised by results. At times, they thought they had a good month, but the results did not support that perspective. At other times, they thought the month was not going very well and found that the closing showed otherwise. They also conveyed that they typically learned of an issue impacting profitability weeks or even longer after the issue had its adverse impact. They indicated that had the issue been known earlier, they could have responded much faster and minimized the negative impact. Many executives related that they felt this was no way to run a business. In fact, a number of executives indicated that they felt their operations were well controlled, while their profitability appeared to be out of control.

At the beginning of this project, we got the impression that the frustration being expressed by the executives had existed for many decades. But on closer analysis, we found that the executives actually believed that this was a fairly recent phenomenon. A number of the executives interviewed who had been with their operations for extended periods actually said that 10 years ago this was not the case. Back then, the operations were well controlled, and profitability was reasonably predictable. Something had changed over the last decade that was throwing the business of industry into turmoil.

Evaluating what may be different today as compared to 10 years ago, we started to focus in on some critical variables associated with the profitability of industrial operations that had been in flux. The first one we noticed was the price paid for electricity. From a stable commodity, electricity has transformed into a business essential that can undergo large, rapid variations in price, with a corresponding great impact on profitability. Electricity pricing was only the first domino in a chain reaction of variations that impacted industrial enterprises in a similar manner. The production of natural gas typically involves significant amounts of electricity. Since the price natural gas producers were paying for electricity was fluctuating at unprecedented rates, eventually, the price of natural gas began similar frequent fluctuations. Raw materials used in industrial operations were another critical business essential linked to the profitability of the enterprise that began experiencing real-time variability.

This shift from highly stable inputs to an operating enterprise over extended periods to real-time variability of these same inputs caused managers of industrial businesses to lose control of the profitability of their business. Profitability became completely unpredictable. Viewing inputs to the enterprise separately from operations, executives concluded their profitability was out of control, while their operations were in control.

It is important to understand that the solution to this issue is not merely to measure the business in real time and then provide those real-time measures to the business executives of the company in a similar manner to the monthly measures. The volume and frequency of the real-time information would quickly overwhelm management. The key is to move critical decision rights down the organization: in many cases, right to the front-line operations and maintenance personnel. This requires providing those traditional functions with the real-time business information in the right time frame to make good business decisions contextualized to their function in the operation. That is, industrial companies must consider changing their traditional "laborers" into "performance managers." This fundamental change to the working culture is an absolutely essential aspect to the success of modern Operational Excellence systems.

This transition of perspective on the front-line personnel from laborers to performance managers has proven to be one of the most challenging stumbling blocks in the quest for true Operational Excellence. It is truly anticultural in many industrial companies. The perspective that front-line personnel are laborers who do not offer much value to the operations in terms of brainpower is a direct consequence of the Industrial Revolution. At this time, a large uneducated and unskilled labor force had to be closely watched and managed, because they could not be trusted to make any operational decisions. Engineers in industrial companies have worked for decades essentially to protect the operation from these laborers by limiting their decision rights and clamping their degrees of freedom. Some of these limitations were required for safety, but many were developed due to a lack of trust of the labor force's ability to adapt and control anything beyond the production process.

However, today's operators and maintenance personnel are vastly different from their predecessors. They tend to have much higher levels of education than ever before. And they are comfortable with views of the production processes through computer-based automation systems that have only been possible for the last decade or so. The result is a well-educated work force with high levels of experiential knowledge gained by the expanded real-time view of the process not unlike video games in which many excel. Continuing to treat such a valuable resource base as traditional laborers is stifling the value they can drive into the business. These valuable resources must become the performance managers of industrial operations. Cultural transitions of this type are typically the most challenging to overcome and are well addressed within the Operational Excellence architecture described in this book.

Moving traditional "laborers" into "performance managers" requires major changes within the organization itself. Agility to accommodate variations in the operating environment quickly demands an end or, at least, a significant modification to the siloed, function-based organization. People throughout the enterprise must work together effectively in teams and essentially in real time to achieve optimal enterprise business/mission objectives. Within this new organizational construct, there must be a much greater awareness of enterprise business/mission objectives, as well as the commitment, ownership, and information absolutely necessary to achieve these objectives safely and sustainably. Perhaps more important, the people themselves have to move into the new order. Human performance excellence demands individuals who are ethical, competent, highly motivated, capable of working effectively in teams, and committed to excellence, continuous improvement, and enterprise business/mission objectives.

Executives likewise have an expanded role in the new model of Operational Excellence; however, in this case, it may be seen as a return to the past. Back to the future so to speak! In the past, it wasn't unusual for the senior executive of an operating enterprise to have risen from the operating level possessing total familiarity with the details of production. These senior executives had worked with and knew many of the operating people by first name, frequently walked the facility and commanded respect. Since they intimately knew the production process, working-level people were comfortable sharing problems that were quickly recognized and corrected. Over the years, as enterprises became larger, senior executives became less engaged with operations, more focused on business and financial reports, and less aware of the linkage between the two. Operational Excellence requires executives to realize that they are the pacesetters, the beginning of the working culture. They must possess a compelling vision, focus on the technical and human attributes that make the enterprise successful, and drive Operational Excellence with continuing, real commitment, and personal engagement.

Information closes the circle. Real-time information supplied to executives that enable their identifying potentially adverse business/mission essentials while time is available for reaction and correction. Effectively organized real-time information to management and working-level employees to assure operations and efforts are consistently focused on safely gaining greatest value from operations. It is no longer sufficient to wish for the status quo; everyone from top to bottom must be totally focused and actively engaged toward continuous improvement and how to make tomorrow better than today!

The net result of all of these changes is that the definition of Operational Excellence within operating enterprises must also transition in order to provide the desired result. In this emerging real-time industrial business environment, Operational Excellence must not be limited to improving operational efficiency as it once had. The domain of Operational Excellence must be expanded to include safety, environmental integrity, profitability, good citizenship, risk, reliability, asset integrity, and human performance improvement, as well as operational efficiency. This expanded perspective provides truly daunting challenges to operating enterprises, but challenges that can be met with modern technologies and the full engagement of an evolving talent base across industry.

On first glance, trying to develop an effective Operational Excellence strategy and approach in today's dynamic operating environment may appear to be a conundrum. But as you traverse through the material herein, you will find a path that will lead to world-class performance through Operational Excellence. It is not an easy or short path, but it will be certainly worth the effort.

This book provides a clear roadmap and detailed guidance to gain effective Operational Excellence. It covers the major areas that must be effectively resolved to move operating enterprises to new and higher levels of safety and environmental integrity, efficiency, effectiveness, and profitability. All greater than previously had been possible and essential for success in the operating environment that exists today and will exist into the future.

1

OPERATIONAL EXCELLENCE—THE IMPERATIVE

The term *Operational Excellence* describes the ideal state of an operating enterprise. It has great cachet at the highest executive levels and is being incorporated into the working culture by organizations committed to being the very best in all business, mission, and operating activities. Whether yours is among the great performing enterprises dedicated to remaining great, a good performer aspiring to become great, or one of the large number who know improvements are essential for continuing success, Operational Excellence and an effective Operational Excellence program are musts. This chapter and the following ones will define Operational Excellence concepts, principles, values, and requirements. Why Operational Excellence is a business and operating imperative, how the program and supporting elements must be mastered, and how Operational Excellence is implemented to achieve greatest results are all described in detail.

DEFINITION OF OPERATIONAL EXCELLENCE

Operational Excellence is an ideal descriptive term; in two words it clearly defines applicability and objective:

Operational refers to an operating enterprise, one that uses some physical means to produce and/or deliver a product and comply with a mission. Thus, Operational Excellence applies to a broad spectrum of industries and operating entities:

Operational Excellence: Journey to Creating Sustainable Value, First Edition. John S. Mitchell.

process, production, power generation and distribution, continuous and discrete manufacturing, mining, food and beverage, life sciences, pharma, transportation, including pipeline, rail, and marine, and many more. Operational Excellence is especially well suited for industries/missions where process flow is fixed and predictability, minimal variation, and reliability are key to effectiveness. In addition to industrial organizations, the process and programmatic implementation described in this book are equally applicable to service industries: municipalities (e.g., fresh and waste water treatment), the military, and other types of operating enterprises where mission effectiveness rather than profit may be the governing factor. Principles expressed to achieve mission and organizational effectiveness apply equally to hospitality (hotel), hospital, financial, insurance, and similar firms that may manage physical and/or financial assets to maximize return.

Excellence is a broad objective that must be achieved and sustained to remain successful.

Definitions from Leading Global Enterprises

Chevron, a leading global enterprise defines Operational Excellence this way:

> "Operational Excellence is the systematic management of *safety*, health, environment, *reliability*, and *efficiency* to achieve world class performance."
>
> Chevron Operational Excellence Management System

DuPont, another global leader states the following:

> "Operational Excellence (OE)" is an integrated management system that drives business productivity by applying *proven practices and procedures* in three "foundation blocks:"
>
> *Asset Productivity*
>
> *Capital Effectiveness*
>
> *Operations Risk Management*
>
> DuPont: Delivering Operational Excellence to the Global Market, 2005

Under the heading of Operational Excellence, ExxonMobil stated their commitment in the 2008 annual report to shareholders:

> "Ensuring the *safety* and *reliability* of our operations is fundamental to our *business success* and a critical challenge that ExxonMobil takes on every day."
>
> ExxonMobil 2008 Annual Report

The commitment to Operational Excellence continued in ExxonMobil's 2011 Annual Report:

> "... we seek to *maximize value and improve efficiency*."
>
> ExxonMobil 2011 Annual Report

Emphasis the author

Operational Excellence was featured in ExxonMobil's 2012 Annual Report as one of five competitive advantages. Four full pages were devoted to Operational Excellence describing ExxonMobil's commitment to Operational Excellence culture, systems, and results in 10 specific areas:

- Management leadership, commitment, and accountability
- Risk assessment and management
- Facilities design and construction
- Information/documentation
- Personnel and training
- Operations and maintenance
- Management of change
- Third-party services
- Incident investigation and analysis
- Community awareness and preparedness.

Under the heading "Delivering Profitable Growth" ExxonMobil's 2013 Annual Report repeats their commitment to Operational Excellence in the 10 areas listed previously. In four pages (two of text), ExxonMobil expresses pride in their "culture of excellence" and states " ... a steadfast commitment to improve the reliability and efficiency of our assets continuously, which leads to improved profitability." Furthermore: ... "our commitment to operational excellence ... provides a solid framework to achieve safe and reliable operations."

With leading global enterprises advocating a continuing commitment to Operational Excellence, there can be no doubt of its great importance to operating and financial success.

A Simpler Definition for Operational Excellence

Safely create greatest sustainable value

In this definition, *safely* goes without question and includes health and environmentally sound operation (SHE/EHS). Value must be defined for the specific enterprise. For some, it is profitability and shareholder value, for others, such as a municipality or public transport system, it could be cost-effective customer satisfaction.

This simple definition is all important. All subsequent processes, tasks, and activities within Operational Excellence detailed in this book will flow from and be prioritized by contribution to "*Safely creating sustainable value*."

OPERATIONAL EXCELLENCE EMBRACES EVERYONE IN AN ENTERPRISE

Operational Excellence requires committed leadership, a positive working culture, and collective and individual ownership for excellence and success. All are necessary for real achievement in any endeavor to improve, gain, and continuously sustain greatness. From visionary, engaged executives, success-oriented leadership, and management to ownership, commitment, and responsibility at the working level, Operational Excellence embraces everyone in the enterprise.

> Ambitious, clearly stated organizational objectives, importance, necessity and benefits, as well as organizational and individual responsibilities, must be clearly enumerated and totally understood by all.

Operational Excellence requires robust, reliable control, management and administrative systems, fully coordinated efforts, and ownership at all levels in the organization. This must be reinforced with complete and accurate practices and procedures, continuing learning and a total commitment to continuous, sustainable improvement.

Operational Excellence broadens horizons, consolidates, builds on, and enhances most existing programs while providing linkages and a laser focus on risk reduction and safely increasing business value/mission compliance. It requires thinking well beyond increasing efficiency to improving effectiveness and achieving results that contribute real value to the enterprise.

OPERATIONAL EXCELLENCE IMPROVES EFFICIENCY

For decades, the primary objective of an operating enterprise has been to improve *efficiency*. Operating efficiency, the ability to deliver a product or service with minimum waste, operating, and energy costs, has been considered essential since Henry Ford's first assembly line. Improvements in efficiency, the optimum combination of variables within an operating/manufacturing/service process, have historically translated directly into corresponding improvements in business/mission effectiveness, quality, delivery, and profitability. Total Quality Management, Lean Six Sigma, and the huge resources expended to optimize production control automation, increase efficiency, and provide more detailed business and financial reporting. All are well-known examples that testify to the importance of this area at the highest levels of executive management. Today, functional improvements in manufacturing, back office, supply chain, etc. are no longer sufficient. Real success requires totally coordinated enterprise-wide improvements in effectiveness.

EFFICIENCY AND EFFECTIVENESS

At this point, the all-important distinction between efficiency and effectiveness must be defined, Figure 1.1:

Efficiency: *performing a given task well*
Effectiveness: *performing the correct task efficiently*

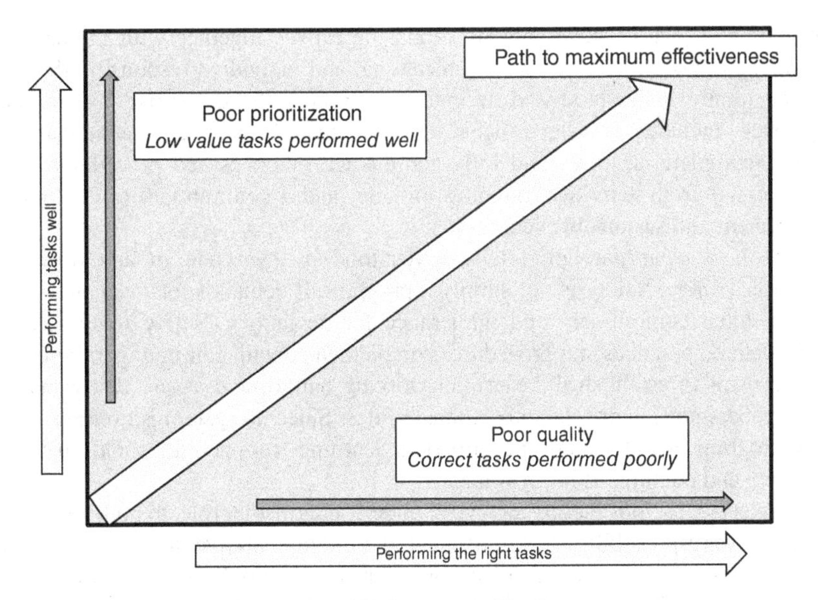

Figure 1.1 Efficiency and effectiveness.

Efficiency is task oriented; it does not question whether the task was appropriate or even necessary. Effectiveness is results oriented; it addresses both the capacity of the task to achieve desired results and how well it was performed. Operational Excellence is focused on results, safe contribution to value—*effectiveness*.

By going beyond efficiency to concentrate on effectiveness, Operational Excellence elevates performance from simply performing activities safely and well, to safely performing the right activities exceptionally well at the right time. This is central to creating strategic advantage and the driver for continuous improvement. High performance organizations apply operational excellence to:

- Create a strategic, enterprise-wide imperative that coordinates and *fully* integrates business, technology, process, and practice for maximum effectiveness, leverage, and economy of scale to gain business/mission success.
- Direct the working culture of the entire enterprise to excellence and continuous improvement.
- Involve and align all relevant functions, processes, activities, and interactions to mutual support and success, including those that are external to the enterprise: interactions with customers, suppliers, and the surrounding population.
- Provide a driving force and catalyst for sustaining gains.

A FAMILIAR PROGRAM

Operational Excellence must include and touch everyone in an enterprise. Operational Excellence typically begins with one or more committed executives conveying ambitious and clearly stated organizational objectives together with importance, necessity, and benefits as well as organizational and individual responsibilities. All must be totally understood and accepted throughout the enterprise. Operational Excellence includes robust, reliable control, management and administrative systems, complete, accurate and fully coordinated practices and procedures, constant learning from activities, ongoing training, and a commitment to continuous improvement and sustainable success.

There is a clear parallel, totally understood by everyone in any operating enterprise—*safety*. Safety is not simply a program. It demands total organizational and individual commitment and intolerance for deviations. Safety requires thoroughly defined practices and procedures complemented with continuing training and reinforcement to establish the essential working culture and assure that everyone clearly understands their role and responsibilities. Since no system is ever perfect, a safety program includes constant reminders, learning from activities and mistakes, follow-up, and continuous improvement.

In a safety-conscious facility, everyone understands his/her role, in both individual and collective responsibility. It isn't unusual for an administrative assistant to tactfully suggest to an embarrassed senior manager that he/she should hold on to handrails while walking stairs. To further illustrate how the safety culture permeates activities, it also isn't unusual to hear safety messages in an industrial facility addressing automobile and home safety. The same must be true for Operational Excellence; it is not a project with a beginning and an end; it is the way for the working life and culture that provides the same positive influence to off work activities as well. Operational Excellence is based on continuously increasing effectiveness: activities that safely and effectively create greatest sustainable value.

> At the end of a review meeting, the plant manager stated: "We must attain a cultural and organizational commitment to operating and asset performance equal to the commitment to excellence we have achieved in the Safety, Health, and Environmental areas."

DESCRIPTION

As will be explained in greater detail in the next chapter, Operational Excellence is generally directed to performance efficiency/effectiveness/profitability depending on specific business/mission requirements. Operational Excellence is the master improvement program that provides a single charter to assure that improvement efforts build up on one another, are optimally coordinated, and develop maximum results. This is particularly important when functional improvement processes overlap.

Operational Excellence is based on and demands a working culture of honesty, integrity, commitment, initiative, ownership and responsibility throughout the enterprise. It is directed to optimizing processes and technology, people, and behaviors. Operational Excellence is largely continuous and evolutionary improvement although it may include step change when there are immediate, large opportunities for improvement.

For many operating enterprises, the long-term future, including future regulatory requirements, competitive, and market environments can't be predicted with any degree of certainty much beyond 3–5 years. Thus, Operational Excellence is constructed around intermediate objectives, waypoints that can be established with certainty along with essential performance objectives. As time moves forward, and future requirements become clearer, objectives and plans are adjusted, refined, and extended along with the performance objectives, tasks, and activities necessary to continue success.

Many enterprises capable of gaining from Operational Excellence are commodity and service businesses where success is gained by quality, response, process, and human effectiveness. In this area, lower cost producers will be assured of continuing prosperity during a downturn.

Growth in top line revenue and market share are sometimes associated with Operational Excellence. While optimizing flow can increase capacity and the opportunity for top line revenue, this is often a strategic executive issue that is out-side the control of the operational level for the following reasons:

- Operational span of control does not extend to market demand, pricing, features, salability, sales, and marketing (ability to locate and close customers), although these factors must be considered and accommodated in forecast operating plans.
- Revenue likewise depends on a number of factors outside the control of operations, such as global economic and market conditions, competitive climate, etc.

Product development is not typically included within Operational Excellence, although many of the principles are applicable. Product development relies on creative identification of future market needs, intense market knowledge, ability to anticipate trends in technology, and often the necessity to create demand for an entirely new concept.

THE JOURNEY

Operational Excellence isn't a project with a beginning and an end. It is a continuing journey of improvement that positions an enterprise to progress constantly forward to excellence and prosperity in the areas most critical for gaining and sustaining business/mission success.

Operational Excellence requires time and a major commitment by all in the enterprise. Foremost is a work culture concentrated on safety, value, excellence, integrity, and continuous improvement. Successful implementation of Operational Excellence as a strategic business/mission essential will generate significant advantages and better financial results compared to enterprises that fail to see relevance or necessity.

Every enterprise embarking on Operational Excellence will have a different starting point, different set of conditions, and different objectives, strategies, and opportunities for improvement. At a high level, it is most important to define business/mission objectives and the scope of the journey itself, which may be at the enterprise, plant/facility, area/unit, or even system/component level. With objectives and scope determined, instilling a continuous improvement culture, empowering people with decision rights, and providing all information necessary are the initial steps in the process. Ensuring alignment of business/mission and program strategies across objectives, actions, and metrics are all important factors to achieve a successful journey.

Roadmap to Operational Excellence

One of the first questions to be asked in constructing the roadmap for Operational Excellence is what is important to the business/mission. This should be spelled out in the business/mission strategy and objectives. It forms the basis for all that will follow while developing the Operational Excellence program. Locally, it is imperative to determine what is important to customers and the surrounding community if not defined in the enterprise business/mission strategy.

As stated, Operational Excellence is directed to seeking safe sustainable value. It includes improving reliability, reducing risk, and variation. What are the key processes that create value for the enterprise; what improvements can be made? How should increased value be measured, and what are the metrics that will be understood by all and promote enthusiasm and ownership for success? How will results and contribution to business/mission success be monitored, validated, and reported?

All these and more will be discussed in the following chapters.

RELIABILITY

Reliability is a term commonly mentioned in the context of Operational Excellence. To many in an operating environment, reliability will be thought of in terms of process and production systems and equipment reliability, even in one's automobile and home appliances. Within Operational Excellence, reliability has a much broader meaning. Reliability is applied to:

- Performance—safely meeting requirements, predictable, minimum variation from best performance (the latter is especially important for establishing objectives and identifying opportunities as will be explained in detail).
- Organization—roles and responsibilities completely defined and understood, consistent decision process.
- Working culture—commitment to and ownership for excellence, highest quality performance, continuous improvement, empowered employees who accept responsibility and accountability for activities.
- Processes, practices, and procedures—completely defined and accurately documented, totally repeatable, high quality, consistent results; method for maintaining currency, implementing, and documenting improvements.
- Systems and equipment—fully capable of meeting all operational requirements safely and cost effectively.

- Skills—requirements and qualifications totally defined, up-to-date, effective training, and follow-up to assure proficiency.
- Data, information, documentation—accurate, secure, up-to-date, and accessible

All these and more will be discussed in detail as vital elements of Operational Excellence.

RISK

Risk is yet another key element of Operational Excellence. It is fully used in the value equation that directs improvement initiatives. What are the probability and consequences of an event that may not have happened that will initiate and justify actions and investment for early detection, avoidance, reduction, and mitigation? Probability has another application within Operational Excellence: assurance that a given activity or task will achieve expected results. Risk, applied within Operational Excellence, will be explored in detail in Chapter 12.

CHANGES IN THE BUSINESS/MISSION ENVIRONMENT

Safely maximizing business and mission value delivered is the sole objective of Operational Excellence. With globalization, and the worldwide integration of many industries, an entirely new way of looking at the connection between operations and business excellence is required. Up until just a few years ago, the business variables associated with most industrial operations had been highly stable and very predictable over extended periods. Costs of energy and raw materials didn't change for months at a time. Two decades ago, electric power was highly regulated. It was not unusual for industrial operating companies to develop contracts with electricity suppliers for periods of 6 months or even 1 year. Electricity was a fixed cost over the contract period. With relative price stability, reductions in electricity consumption directly translated into predictably lower energy costs. Alternate sources and methods, for example, waste steam, compared to electrically powered equipment within a process facility could be financially evaluated with certainty. Other key business operating variables such as material and other utility costs and product market price were similarly stable over a reasonable planning period.

In the current operating environment, stability has been replaced by uncertainty. Electric power costs can change by the minute. Raw material prices fluctuate in response to global demand and political conditions. Government regulations add cost and create uncertainty. Practices that may have been totally acceptable in the past may be ruled impermissible, thereby incurring large costs for compliance over a relatively short period.

Globalization creates huge disparities that must be overcome. How does an enterprise produce a competitive, sustainable, and attractive business value within an industry where global labor costs might vary by an order of magnitude for equivalent quality? Furthermore, production capacity can emerge quickly anywhere in the world with a lower cost structure and the ability to deliver products meeting all commercial specifications. Whether we like it or not, we are now in a real-time business environment. It may be necessary to instantaneously alter operating decisions made in the past under one set of conditions that are suddenly invalidated by changes over which the enterprise has little or no control. The concept of keeping an operating enterprise "evergreen" now requires continuous review, reassessment, and improvement; all addressed by Operational Excellence.

CONVENTIONAL OPERATIONS MANAGEMENT

Most operating enterprises are organized by function: sales, marketing, operations, maintenance, engineering, finance, information technology (IT), human resources (HR), etc. that are managed independently through the senior executive level. Functions are typically islands of competency ruled by jealous kings, populated by antagonistic armies, and separated by shark-filled seas. Typically, there is little structural or organizational encouragement for groups to work across functional lines. In many process and manufacturing enterprises, operations doesn't like maintenance, neither like finance or IT, and the feelings are reciprocated! At the operating level of a production enterprise, conflicts between operations and maintenance are common and diminish overall effectiveness. An ever-increasing operating tempo and cost of downtime can cause even greater friction.

Complicating matters, production operations are typically viewed as the profit-making portion of a manufacturing enterprise, with performance objectives primarily based on throughput. Within the same enterprise, maintenance is commonly viewed as a service, a cost measured by availability and compliance to budget. The profit center, cost center disconnect may appear small but is actually huge in terms of the working culture where throughput, asset availability, and maintenance cost are often inverse functions. Increasing throughput, running a facility, process, or system harder to produce more output generally increases maintenance costs and may decrease asset availability.

As a further complication, differing requirements between business and operations management have resulted in separate business and plant level management systems. Business management systems are typically backward looking transactional accounting systems. They are managed by finance and IT departments to meet requirements for financial management and reporting. Systems report results of events and decisions made months, often even years in the past.

Operations management focuses on real-time process control, production output, and efficiency. Operations management and support systems are designed and operated from an engineering-based operational perspective. There is very little business, mission, and profitability context. Thus, information of prime importance to one function may not be as important in the view of another. Into this often disconnected structure add additional islands of information designed for a specific purpose within a single function. These disconnects can be a significant weakness, especially as industrial businesses are transitioning to more real-time dynamics. More about this is in Chapter 6.

Within manufacturing enterprises, the path to greater efficiency has typically occurred in two separate paths:

- Process and control automation has advanced rapidly reaching today's level of sophistication to maximize efficiency in approximately 25 years.
- Huge investments have been made in business systems, primarily to integrate and automate the financial management and accounting processes. This is the transactional reporting system mentioned earlier.

While many recognize the necessity, an equivalent organizational commitment to improving the performance, efficiency, and reliability of the physical systems, and assets (structures, vessels, heat exchangers, piping, conveyors, machine tools, fixed and rotating equipment, etc.) on which production and production effectiveness are absolutely dependent has proceeded at a much slower pace. Advances in management of physical assets has lagged and been largely disconnected from those in the process control and financial areas.

MAINTENANCE WITHIN AN OPERATING ENTERPRISE

Within a typical production/operations mindset, the basic idea of maintenance has remained unchanged since the industrial revolution. The physical plant is expected to perform; maintenance occurs when it doesn't. Maintenance has been considered a service and a cost to be controlled rather than the integral part of enterprise profitability/mission compliance it should occupy. Availability is treated as an average rather than a potentially sudden, unexpected event that can, and often does, impact production delivery in the worst possible way.

Only the most enlightened enterprises consider maintenance as an essential component of the core business value producing process, a fully empowered, equal partner of production operations. A growing realization of importance has resulted in the release of ISO55000, Asset Management that will gain maintenance an equivalency to ISO9000, 9001 Quality Management. Operational Excellence solidifies the relationship and moves the enterprise working culture to a most effective partnership.

MANAGING IMPROVEMENT INITIATIVES

For the most part, improvement initiatives within an operating enterprise have been function specific and managed as independent programs rather than an overall coordinated process. For example, many industrial companies initiated specific programs within a function for energy management, production throughput, quality, work and material management, and personnel efficiency. Safety and environmental improvement programs were the exception that crossed functional lines.

Each functional program was typically headed by a separate executive, such as an Energy Czar, with independent teams working on each initiative. This programmatic approach is based on an underlying assumption that process and system improvements are reasonably independent and that improvements in one area will not have an appreciable impact on any of the other areas.

This assumption is generally incorrect. Consider an energy management program with the primary objective of reducing energy consumption. The best way to meet this objective would be to turn off the operation. This would obviously meet the objective of the energy management team. Unfortunately, all would lose their jobs, as it would create a negative net business value. Furthermore, turning off the operation makes it a bit more difficult for the team charged with maximizing production throughput to achieve their objective. In this case, and in most others, tradeoffs are required to assure that the overall business/mission system is optimized.

Efforts to improve one or more business or operating parameters often result in an abundance of uncoordinated and disconnected initiatives. Within an operating enterprise, there is generally a very close relationship between throughput/output, quality, energy consumption, material consumption, safety and environmental integrity, and cost. As cited previously, well thought out but uncoordinated improvements in one area can affect another area with unintended consequences that may include a decrease in the overall value produced.

Process operations within a batch manufacturing facility initiated a Six Sigma improvement project to reduce Work In Process (WIP). Without consulting maintenance for an assessment of complications that might occur with increased product changeovers, the decision was made to reduce batch quantities to single orders rather than forecast demand. The resulting increase in changeovers from one batch to the next reduced production availability as well as increased maintenance costs for the cleaning and equipment changes required between batches. Overall result was a net negative.

With an abundance of isolated initiatives and often conflicting objectives, it's little wonder that functional teams haven't tended to work well together. Working-level employees within a series of uncoordinated and disconnected initiatives lacking an overall vision or strategy generally conclude that the latest initiative is just one more "program of the month" to be ignored until the fellows in the corner offices lose interest as they always have in the past.

A large corporation launched a series of functional initiatives to standardize and improve performance. A participant in one initiative accidently recognized a similar initiative being pursued in another functional area. The corporate CEO was the only common organizational link between the two initiatives! Efforts to coordinate the initiatives at the working level favored by members of both teams were initially discouraged and then forbidden as "out of scope." A great potential to assure optimum, coordinated, and fully aligned improvements in two functional areas was lost.

Despite all the talk about organizational collaboration over the past decade, it is discouraging to witness examples such as cited previously. In many cases, management decisions, performance measurement systems, and organizational and individual objectives may actually reduce cooperation and produce suboptimal results when viewed from an overall value perspective.

> A worker commented that while expecting/prioritizing the necessity for new behaviors and demanding performance in value- and result-oriented ways, the organization remained static and disconnected with performance driven and evaluated by old activity-based metrics. Management always appeared surprised when a new initiative did not produce any change. "Why should anyone be surprised when success is always evaluated with the same old measures of performance?"

It should be clear that taking a traditional function-based programmatic approach to operating efficiency will seldom result in optimal overall effectiveness. Producing greatest value requires balancing all variables to optimize business, mission, and operational objectives. Coordination and cooperation across organizational boundaries nearly always produce greater results than function-specific initiatives.

It may be acceptable and desirable to increase energy consumption if increased production throughput/mission compliance results in greater overall value gain, including increased energy and maintenance costs, without diminishing safety or environmental performance. Exceptional care and a thorough analysis must precede this conclusion. An accurate business/financial model, described in detail in Chapter 5, is essential to assure that a real increase in value is delivered.

There has to be a better way—and there is.

THE SOLUTION

With its continuing focus on "*Safely creating greatest sustainable value*," the concept of Operational Excellence is the bright star in the sky. Operational Excellence is an overall master program addressing all issues with a value improvement objective. Rather than replacing successful practices and programs, Operational Excellence knits them into a larger coordinated and fully integrated tapestry constructed to improve value produced within the enterprise business/mission strategy. Think of Operational Excellence as the roof under which all functional improvement programs live, are coordinated, and thrive, as illustrated in Figure 1.2. The Operational Excellence program is built around the necessity and benefits of working cooperatively across functional barriers. It includes complementary, mutually reinforcing, internal processes, as well as a time horizon and response mechanisms that are short enough to assure continuing success within a changing operating environment.

> Difficult to see revolutions when they are happening—occurs only with hindsight!
>
> Dr. Peter G. Martin Vice President; Invensys

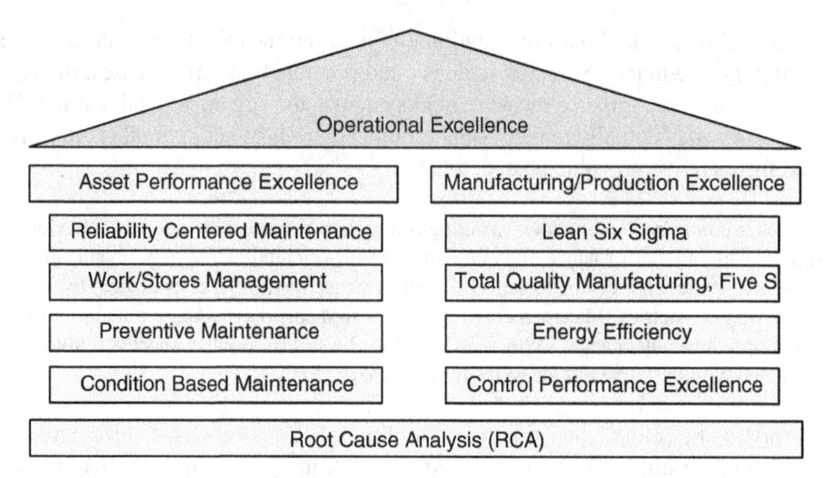

Figure 1.2 Operational Excellence—the roof housing all improvement programs.

Leadership

Will people at the working levels understand and embrace Operational Excellence and recognize its potential to gain the latent effectiveness so many know is available and on which their job security and compensation may depend? A large part of the answer revolves around leadership and communication. Senior executives and operating leaders must be totally committed and convey a vivid picture of why Operational Excellence is essential. Through personal example and reinforcing communications, leaders must make it clear that Operational Excellence is a continuing business/mission imperative, not a passing "fancy of the month." They must establish clear objectives, lead, and drive the process with visible, active, continuing personal engagement, and provide incentives for success. Leaders must emphasize the necessity to gain results that increase business/mission value along with the supporting activities necessary to sustain success.

> High energy, passionate, driven leaders setting a personal example and continually demanding excellence and improvement will energize those whom they lead and influence the work force to achieve results they might not imagine possible.

Operational Excellence provides the process, system, organization, and methodology to address and assure success in all these areas.

Working Level Improvement Action Teams

Operational Excellence depends on results produced by multifunction, working-level action teams directed to develop and implement improvements that safely increase value and reduce risk. One step above, and best led by a production manager, leadership teams identify and value prioritize potential improvements. Leadership teams appoint action teams with the experience and skills necessary to develop, implement, and monitor results of value improvement initiatives. This process will be described in more detail in Chapter 7.

During a brief meeting to summarize requirements and benefits of Operational Excellence, one of the participants concluded that the idea of production leading Operational Excellence improvement action teams implied that Production would take over Maintenance, an action with which he totally disagreed. He was reassured that Production leading Operational Excellence action teams did not in any way mean that Production was taking over Maintenance. Maintenance remained an independently managed function with responsibilities for contributing to the improvement action team process and implementing improvement actions as appropriate. Joint participation in action teams established and reinforced the essential notion of an operations/maintenance partnership that is essential to gain maximum success.

Combining operations, maintenance, engineering, finance, IT, HR, and others into a team with a common purpose to increase value produced quickly breaks down functional barriers. Team members rapidly identify and accommodate varying perspectives and learn to devise mutually beneficial value improvement initiatives. Of all the elements of Operational Excellence, working cooperatively across functional boundaries to identify and develop improvements is by far the most important.

"Employee led leadership teams do unbelievably good strategic and tactical planning—if you give them the opportunity. Most important, they gain total buy-in for the plan, its implementation and results."

Retired Fortune 250 company CEO

Operational Excellence Improvement Initiatives

As an enterprise-wide initiative with everyone involved and participating, Operational Excellence has many advantages over function-specific improvement initiatives. Although there are exceptions, history indicates that the latter generally never reach sustaining. Function-specific improvement initiatives are typically implemented to solve a problem or problems; low quality production, too many failures, too much work in process, availability less than required, excessive costs, poor quality work, etc. The initiative succeeds, problems are solved, victory is declared, emphasis is directed elsewhere, and the reason for the initiative fades into the forgotten past.

Next comes a profit challenge, from pricing pressures, reduced sales, change in the competitive environment, more capacity coming on line in the market, etc. First reaction is to cut costs. Looking for potential savings, function-specific programs are scrutinized: "why does this program exist? Haven't had any failures in memory, everything runs well, quality is great, cancel the program, and either lay off the participants or shift them to work in another 'more productive' position." Eventually, initial conditions return and no one recalls why. The cycle repeats anew.

A long time salesman commented to a plant person appointed to head an improvement program that this was the third time he had sold identical program components

> to the facility over about 10 years. Knowing that the efforts were initially successful, he asked: "What happened to the first two efforts?" The individual stated there were no records or institutional memory indicating the program had been implemented in the past.

Operational Excellence involves everyone in the organization. It broadens horizons and builds on, consolidates, and enhances most existing programs while providing linkages and a greater focus on the whole. It requires thinking well beyond increasing efficiency to improving effectiveness, thereby achieving results that contribute real value to organizational objectives in real time. The Operational Excellence program's structure and internal culture sustain themselves.

Sustainability

Gaining results through consistent, sustainable practices is an essential element of Operational Excellence. Some enterprises have focused on performance objectives without assuring that the practices and procedures necessary to attain the improvements are mutually supportive, institutionalized, and sustained. The requirement for sustaining improvements is exceptionally important to the success of Operational Excellence. Without discipline and a commitment to sustainability, it is very tempting to prematurely declare victory and terminate an improvement initiative that has attained most of its objectives. Within Operational Excellence, programs and improvement initiatives continue until all elements that have produced success are sustainably embedded in the working culture to the point it is simply "the way things are done."

EFFECTIVENESS AND VALUE THROUGHOUT THE ENTERPRISE

An operating enterprise requires a multiplicity of activities and tasks to make up the overall business/mission tapestry. At the working level, activities and tasks justify the job. Performance is thus hugely important to the person with direct responsibility. As important as they may be to an individual employee, there is typically little linkage between performance of individual working-level activities and tasks to business/mission effectiveness and value.

The same is true in reverse. Just as the picture in a complex jigsaw puzzle can't be visualized from a few pieces, a financial executive concerned with cost effectiveness, profitability, and shareholder value who frets constantly over business results can't connect contribution to corporate value from activity/task performance effectiveness at the working level. Operational Excellence provides the bidirectional awareness, linkages, and sight lines to translate practices, activities, and tasks into business/mission effectiveness and value. It will assure decisions at all levels are complementary and contribute to the overall objective: *safely produce greatest sustainable value.*

Value prioritization is a prime factor that must be uppermost in mind. Which activities and tasks contribute most to value and business success and how? There are always more opportunities for improvement than time and resources. How are available time and resources used most effectively? What is the sight line between a given task/activity and value to the business/mission? If it isn't there, can't be defined, has low value-add or low probability of success the task/activity should be reconsidered, modified to create acceptable value, or perhaps eliminated altogether. And that leads to another question: can business results be controlled in real time? The issue is beginning to be addressed by enterprises where agility to meet changes in their competitive environment can spell the difference between success and failure. This issue is discussed in more detail in Chapter 6.

THE OPERATIONAL EXCELLENCE INITIATIVE

With the necessity established, the next question is where is Operational Excellence on the road to general acceptance? Some well-known enterprises mentioned at the beginning of this chapter are committed to Operational Excellence, with relatively well-defined processes for implementation. On the whole, Operational Excellence is in its early stages with great promise. Solid, universally accepted definitions and an implementable program are yet to be fully defined. The wide range of operating enterprises to which Operational Excellence is applicable makes it more difficult to establish solid definitions and implementing processes. This challenge is discussed in detail in the next chapter.

Many corporate executives have concluded that Operational Excellence is a good idea and are seeking more detail and looking for implementing methodology. Many more at the middle level of organizations have been charged with implementing Operational Excellence without a solid idea of how to translate an executive directive into an implementable program that will attain the results demanded. This book will fill that gap.

Developing and implementing an Operational Excellence strategy across multiple functions within an operating enterprise can be highly complex and extremely daunting. Dividing the program into focused areas under a comprehensive strategy, with clear objectives, robust leadership structure, and a well-defined implementing process has major advantages. It builds from a functional organization that all in the enterprise are familiar and presumably comfortable. It reduces complexity and engages the talents of the enterprise most effectively. Most importantly, it keeps the people affected reasonably within their comfort zone, while opening the opportunity for major improvements and greatly improving the probability of success.

Operational Excellence within an operating organization can be subdivided into three major groups. Figure 1.3 expands and refines Figure 1.2:

- Operating performance excellence: optimize production scheduling/mission compliance, flow, process, and conversion effectiveness.
- Asset performance excellence: maximize the effectiveness of production assets.
- Support performance excellence: optimal effectiveness within engineering, finance, HR, etc.

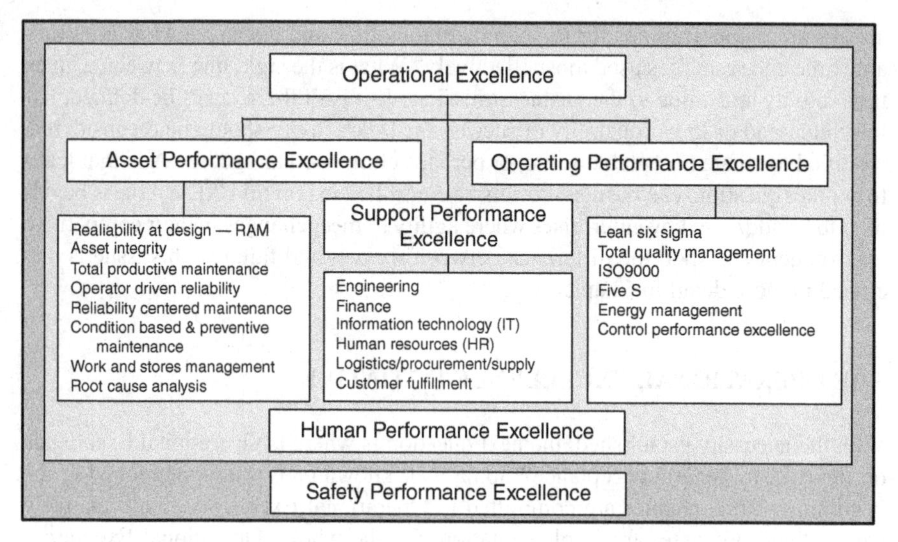

Figure 1.3 Functional divisions within Operational Excellence.

Business systems, supply chain, and sales and marketing are included in support performance effectiveness and use the same principles and processes.

Safety and human performance excellence apply equally to all.

This grouping will be expanded in later chapters.

Operating Performance Excellence

Operating performance excellence is directed to optimizing production flow, scheduling, efficiency, and process parameters for most efficient operation and mission compliance. Process parameters include variables such as pressures, temperatures, speeds, tolerances, etc. Total Quality Manufacturing (TQM) and Lean Six Sigma are examples of typical proven practices. These are reinforced with a variety of process automation optimizing algorithms and procedures including Control Performance Excellence to be discussed in Chapter 6.

Asset Performance Excellence

Asset performance excellence is defined as follows:

> *Safely* gain optimum *sustainable* lifetime value, utilization, productivity, and effectiveness *from physical manufacturing, production, and infrastructure assets.*

With this broad outline, recall that Operational Excellence is an overall holistic initiative. Therefore, all elements are managed for consistency, coordination, cooperation, mutual reinforcement, and contribution to the value objective. There will be a great deal of cross-function coordination to assure development of optimum improvement initiative. The Operational Excellence Steering Team, to be introduced in Chapter 9, has primary responsibility for coordinating cross-function activities to assure greatest sustainable value.

The ISO55000 series mentioned earlier specifies the strategy, management system, and controls necessary to create greatest value from operational assets. Because asset management is an integral subset of Operational Excellence, it is imperative for the Operational Excellence strategy, management, and control structure to be fully compliant with ISO55000.

SUCCESS—GREATER THAN THE SUM OF THE PARTS

As illustrated in Figure 1.3, operating and asset performance excellence along with other functional improvement programs all exist under the roof of Operational Excellence and are mutually complementary. Cross-function coordination and cooperation, overseen by the Steering Team, is a large, essential part of gaining optimum results in all areas. Everyone has mutual responsibilities, and all elements mesh together seamlessly toward a common purpose. Within Operational Excellence, Production Operations, Maintenance, Engineering, Information Technology, Human Resources, Finance all work together cooperatively from one menu to assure that the appetizers compliment the main course. Each person concentrating on their responsibilities is totally confident in the knowledge that everyone else is doing their part to make all the elements complementary and highly successful. The following abbreviated list will be detailed in later chapters.

Essentials for Success

- Active, visibly committed leadership at all levels, continuing reinforcement from the executive and senior management level
- Enterprise business/mission objectives and strategy fully defined
- Operational Excellence program detailed by a written charter including clear objectives, organization, plan, milestones, and metrics
- Program strategy and objectives harmonized with business/mission strategy and objectives to assure consistency, efficiency, and maximum contribution
- Organizational structure designed for effectiveness and sustainable success
- Commitment to excellence, continuous improvement throughout
- Engaged, trusting, committed, responsible, ownership working culture in place
- Multifunction teams at the working level concentrating on attaining optimal performance, eliminating deficiencies, and empowered to implement improvements
- Improvements implemented considering value and time to achieve gains
- Quick wins with real benefits to demonstrate Operational Excellence works
- Continuing communications to publicize successes, establish, and build support

- Assure basics are in place: operating and work practices and procedures fully detailed, followed, and used for training
- Accurate, secure, and user-friendly data and information systems
- Continuous follow-up.

APPLICATION

This book concentrates on value-driven Operational Excellence. Because the application and implementation of Operational Excellence to production efficiency and most of the other functional areas are very specific to the operation and mission, many of the examples, discussions, and explanations will be based on Operational Excellence applied within manufacturing/production enterprises broadly defined to include discrete manufacturing, process, and production companies, as will be explained in the next chapter. Although categories and details may differ, the same basic concepts hold true for other mission-oriented and service enterprises. The concepts, organization, human elements, and methodology are identical for all implementations of Operational Excellence.

WHAT YOU SHOULD TAKE AWAY

Operational Excellence is the single, master, unifying improvement program for the enterprise. It is the enterprise-wide improvement program for production and business system effectiveness and increased value. Production/mission efficiency, asset management, etc. all fall under Operational Excellence. The Operational Excellence program is equivalent to safety in terms of procedure, organizational, and individual commitment. Common objectives, unified management, control and administrative systems, and coordination and active communications between functional initiatives assure complementary, mutually reinforcing activities. On the basis of multifunction improvement action teams, requirements in one functional area that may require participation by another are immediately identified and resolved. Finally, successes and lessons learned are communicated throughout the enterprise by the Operational Excellence program so that everyone benefits.

Advantages and benefits of an opportunity-driven Operational Excellence program to *increase business value safely and sustainably* reduce risk, and lost opportunity includes the following:

- Gains highest safety and environmental performance
- Increases business value and operating effectiveness
- Reduces risk
- Leads to all the processes and improvements necessary to establish and maintain mission/industry best performance
- Results in greatest value, operating effectiveness, and reliability

- Gains optimal resource effectiveness
- Builds an effective organization and working culture
- Leads to improved practices and procedures
- Demonstrates results
- *Sustaining.*

Greater detail will be found in Chapter 4.

2

APPLICATION OF OPERATIONAL EXCELLENCE

Operating enterprises capable of gaining full benefits from Operational Excellence are many and varied. The environment within which an enterprise operates, the enterprise's business/mission strategies, objectives, processes, and operating characteristics of the enterprise in themselves determine how Operational Excellence is developed and applied to gain greatest value and return. One size doesn't fit all; one journey doesn't satisfy all requirements! Formed around principles, Operational Excellence must be custom tailored for the enterprise strategy, specific performance objectives, current levels of effectiveness, and the mission/market itself. Since Operational Excellence typically begins from a necessity to improve an enterprise's value-producing process, this chapter will describe how different process characteristics, short- and long-term tactical and strategic considerations, the mission/market, external environment, and specific requirements for value creation all act to shape an optimum Operational Excellence strategy and program.

PROCESS CHARACTERISTICS

For the purposes of this book, requirements for Operational Excellence are defined in the following three generalized process types:

- *Continuous*, in which process flow is determined by product and is not alterable in any significant way. Industries in this category are characterized by a large

Operational Excellence: Journey to Creating Sustainable Value, First Edition. John S. Mitchell.

concentration of physical assets, typically few inputs (raw materials, utilities, etc.), and a small number of products. A major increase in throughput generally requires an acquisition and/or capital project addition. The category includes oil and gas production and transmission, oil refining, petrochemical, power generation and distribution, water treatment and distribution, paper manufacturing, aluminum and steel smelting, and rolling.

- *Discrete manufacturing* where process flow may be improved to increase throughput, efficiency, and quality. The supply chain, both input (raw materials, parts, subassemblies, utilities, etc.) and output (products) are typically more numerous, varied, and complex compared to the continuous category. The automotive and aerospace industries are the most visible in this category. It also includes numerous manufacturers of precision parts, components, and subassemblies that are sold directly and/or to other discrete manufacturers for incorporation in their products. Process flow in back office, purchasing, sales, marketing, distribution, and supply chain functions can often be improved and are included in this category.
- *Hybrid enterprises* that share both of the preceding characteristics. Batch processing, transportation (marine and rail) industries are examples. Within these groups, physical assets such as process components, ships, trains, and rail systems are in the first category. Operating processes such as scheduling, product changeovers, loading, and unloading are in the second category.

The opportunity to gain significant value by optimizing process flow is a primary difference between the continuous and discrete categories. The first category is asset intensive; flow is dictated by physical assets and a fixed production process and cannot be altered in any substantial way. Within these industries, optimizing reliability and effectiveness of the process is a key value producer. In the second category, namely, discrete manufacturing, office, and supply chain, methodologies such as Lean Six Sigma are in widespread use for optimizing process flow.

The primary element of Operational Excellence is to identify and justify the type and potential of improvement initiatives such as efficiency, reliability, flow optimization/Lean Six Sigma in terms of effectiveness and value gain before embarking on a major improvement program. It also recognizes that increasing production throughput or even introducing an innovative new product may not increase value delivered, unless there is assurance that a market will be available at a profitable price and a solid, credible marketing plan with a high probability of making that happen has been developed. This is a fundamental principle of Operational Excellence and contrasts to the usual implementation of an improvement program: direct attention to a specific process or objective such as increasing efficiency on the assumption that value will be gained because an operating enterprise somewhere else (and perhaps in another segment) gained success. This fundamental advantage of Operational Excellence will be discussed in more detail in Chapter 4.

A large instrumentation manufacturer was about to release a highly innovative new product with significant benefits into an established market. During development of the marketing strategy and plan, the question was asked: how will existing participants in the market respond to a new product that obsoletes their current products? The manager of the new product replied that competitors couldn't do anything faced with such an innovative new product with so many significant benefits—the new product would immediately capture a major market share. Of course, the established suppliers could and did do a lot. They reduced prices, forcing buyers to choose between a very expensive innovative but unproven new product and the proven older generation product. At the same time, all rushed to develop a technical equivalent. Totally unable to profitably answer the competitive response and lacking the deep pockets and market position of the established suppliers, the innovative new product failed to gain any appreciable market share and was eventually withdrawn.

Since flow optimization is a basic difference between categories and there is ample material available on this vital subject, this book concentrates on the application of Operational Excellence to improve value through continuing improvements in efficiency and effectiveness. This vital element of Operational Excellence is applicable across all categories of operating enterprises and demands equal emphasis with flow optimization. When the Operational Excellence value identification process identifies flow as a major opportunity for improvement, readers are referred to books specifically addressing optimizing processes such as Lean Six Sigma.

OPERATING/MARKET ENVIRONMENT

The operating/market environment is another major distinction between operating enterprises. There are at least three general categories, probably more, with distinct characteristics:

- Static, slowly growing, or declining in a predictable manner
- Rapidly expanding and/or changing
- Very large, relatively stable market in which competitors struggle for incremental changes in share without significantly altering the overall environment.

Gasoline, electric power, and toilet paper are examples of manufactured commodities in the first category. Population growth is one of the largest predictors of demand. Changes can occur due to outside economic conditions.

Social media and personal electronics are examples of the second category. It is very difficult to predict the emergence of a market and even more difficult to predict growth, although venture capitalists make every effort and many succeed.

Many manufacturers in the third category supply a market that is so large in terms of units sold that major improvements have only a small effect on the market. The ubiquitous traffic cones and tubes that delineate lanes during road construction are a familiar and highly visible example of a manufactured product in a huge market.

While the value concept and principles of Operational Excellence are valid for all, the preceding distinction is made to demonstrate that "one size fits all" doesn't work for the details of Operational Excellence. Although all may fit under a general definition of enterprises for which Operational Excellence offers many compelling advantages, a program devised for an automobile manufacturer may or may not work for a manufacturer of traffic cones, aircraft wings, automobile seats, electrical power, or an enterprise pumping oil out of the ground or distilling gasoline. Objectives, principles, and implementing details must be weighted, adapted, and applied to fit the specific market/operating environment, business, and mission requirements of the enterprise.

ENTERPRISE STRATEGY

An overall enterprise strategy is absolutely essential to define the "where we must be" question, as illustrated in Figure 2.1. The strategy defines where we must be and the high level actions necessary to get there. This strategic framework provides the basis for an enterprise to make both strategic and tactical decisions that move the institution closer to "there." Since "there" is the enterprise business/mission objective, the Operational Excellence program must be designed to fulfill the long-term strategy. This essential, along with the necessity for the strategic framework to be fully understood and form the basis of improvements throughout the enterprise, will be repeated and emphasized throughout the program implementing details beginning in Chapter 9.

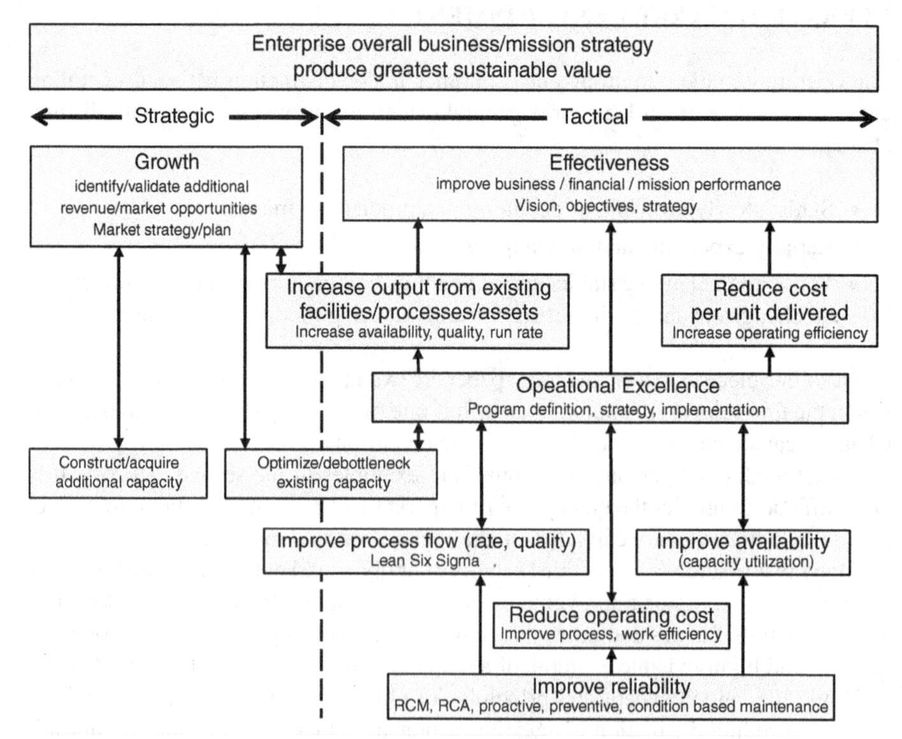

Figure 2.1 Strategic and tactical elements of producing value.

Devising the overall business/mission objectives that will drive an Operational Excellence initiative is a strategic issue determined at the highest level of an operating enterprise. Whether increased production, revenue growth, some other consideration, or a group of considerations dictates the business/mission strategy, it is generally not under the control of any of the direct participants in Operational Excellence. Furthermore, the business/mission fulfillment strategy can change in response to changes in the operating/market environment. Thus, the Operational Excellence program must be constituted with sufficient range and flexibility to provide results within any foreseeable change in the operating/market environment. Typically, this will result in an Operational Excellence program directed to continuously seeking improvements in effectiveness and value delivered.

Strategic and Tactical Endeavors

Referring to Figure 2.1, within the overall business/mission strategy of operating enterprises, there will be strategic and tactical endeavors. Strategic endeavors are typically long-term, forward-looking efforts to grow the business/mission and assure it succeeds within probable changes to market/mission opportunities, operating, and regulatory environments.

Product, business and capacity planning, and marketing are four key functions in the strategic category. The first develops ideas and applications to assure the business remains ahead of and successful within the market/mission mainstream. Business and capacity planning are directed to assuring business results going forward and capacity available to meet forecast requirements. Marketing seeks to identify and develop existing and new markets/missions for the enterprise's current and future offerings. Relative importance and emphasis depend on the specifics of the market/mission.

The tactical side of an enterprise consists of operations, the actual process of most effectively fulfilling an enterprise's business/mission objectives. Manufacturing/production, service delivery, maintenance, the supply chain, and most back office functions are in the tactical side of the enterprise. Sales, which is the continuing effort to assure a customer for the business output, is also a tactical function.

There is actually a third category: support functions. Purchasing, Human Resources, Finance, and IT are examples of this category. Typical support functions were shown in Figure 1.3.

There is another factor to consider. When an individual or an organization has both strategic and tactical responsibilities, the tactical generally gains greater attention. That's because the tactical often demands immediate action; most will conclude the strategic can be postponed. When an operating enterprise combines sales and marketing functions into a single organization, marketing typically has less emphasis than sales due to the necessity of getting immediate business into the door. The same holds true for production. It is difficult to focus attention on strategic areas that participants may conclude have little applicability to their day-to-day duties, responsibilities, and challenges. This tension, present in virtually all operating enterprises, emphasizes the necessity for continuing education and training within Operational Excellence.

Similarly, combining two defined functions under a single manager often has the same, less than optimal results. As one example, combining the operations and maintenance functions under a single manager, generally with a production background, typically demotes maintenance as a service to operations rather than the partnership demanded by Operational Excellence. This will be explored in greater detail in later chapters.

One solution that has successfully ended conflict between operations and maintenance is to switch managers without warning. The ex-operations manager, now maintenance manager, sees first hand the necessity for cooperation. Likewise, the ex-maintenance manager, now operations manager, sees pressures and challenges that previously had not been recognized or fully appreciated.

Growth Strategy

As illustrated in Figure 2.1, there are three methods to grow an enterprise:

- Construct/acquire additional capacity
- Debottleneck existing processes to increase capacity
- Increase output from existing processes through improved effectiveness.

The first, construct/acquire additional capacity is almost always a capital project and not included in Operational Excellence until the capacity comes on line, although the planning process will begin earlier. The second may involve both capital expenditures and process improvements. The third will be almost totally within Operational Excellence as illustrated in Figure 2.1.

Effectiveness

Improving operating effectiveness is the primary objective of Operational Excellence. This can be accomplished by improving production efficiency and/or reducing cost, both addressed by Operational Excellence.

These differing perspectives within an operating enterprise illustrate, emphasize, and reinforce the concept that Operational Excellence will mean many things to many people. Operational Excellence bridges the gap and provides the means for coalescing and mutually reinforcing activities in pursuit of the enterprises value objective. By focusing efforts on business/mission improvements identified through strategic innovation and tactical necessities, Operational Excellence builds consensus, ownership, enthusiasm, support, and sustainable results.

CHANGES IN THE OPERATING ENVIRONMENT

Potential changes to the mission/market, operating, and regulatory environment can force major changes into an operating enterprise and certainly the necessity for and application of Operational Excellence. The emergence of lower cost-competitive capacity has significantly altered the landscape in labor-intensive industries such as textiles, apparel, and consumer products. In some cases, increased automation, production effectiveness, and delivery are insufficient to overcome geographic competitive advantages. Furthermore, it has become very clear that countries primarily producing raw materials have recognized the value advantages of moving downstream into higher value products further eroding the economic viability of more traditional manufacturing areas.

Operational Excellence is a vital ingredient in preserving the manufacturing base essential to economic prosperity in areas of the world with high intrinsic labor costs and strict regulatory requirements. Operational Excellence provides the essential effectiveness and value for these areas to compete successfully in many industrial fields.

Some operating organizations mentioned earlier, power generating, water, and waste water treating, may consider themselves invulnerable to global competition. That's false security. If an industry can't survive profitably, demand for commodities such as electric power diminishes. In addition to reducing economic activity, reducing the demand for a commodity such as electricity also increases the price for all who remain. Much the same is true for water and many other operating industries. Fewer users require spreading fixed costs over a smaller base, with a resulting price increase for those who remain. Failure to produce sustaining value at the new cost structure requires increased prices and in some cases may eventually end the business with corresponding loss of economic activity. That depresses an entire geographic region. Communities and areas decline as real estate values plummet, and there isn't sufficient economic activity to support the necessary retail, service, and medical infrastructure.

> A company party was held in a public restaurant in a depressed area that had seen the shutdown of most of its formerly large manufacturing base due to a combination of the cost of raw materials and electricity, inefficient processes, and price competition from lower cost areas. As the party concluded, a woman asked if participants were from one of the two remaining manufacturers. When one of the participants replied in the affirmative, she stated "you have to do everything possible to keep your company here and in business—this entire region depends on you!"

Although it is unlikely the woman in the preceding paragraph was technically oriented, she knew the economic imperative of keeping the enterprise in business and prospering. Operational Excellence is a vital, indispensible contributor to this essential objective. Incidentally, the company in question is a vocal advocate of Operational Excellence.

VALUE

Value and creating value is the essential element that must be addressed in the Operational Excellence process. For some enterprises, greatest shareholder value is gained through increasing market share essentially independent of profitability. This is primarily a strategic process with tactical elements. High technology consumer goods and social media are the best examples. A new idea often gains market acceptance and shareholder value on the basis of perceptions of the future. For enterprises in this category, top line revenue growth and staying ahead of potential competition are typically more important than profitability. This works so long as there are investors willing to bet on the future and a credible plan to gain profitability before losing advantages of growth and market share.

Most continuous and many discrete industries operate within a reasonably predictable market, and many in commodity markets. Although sales and marketing spend a great deal of time constructing a differentiating message, there is typically little real difference between the product offerings of the top three or four enterprises in a commodity market. Differentiation is achieved by excellence in areas such as market presence, response, delivery, and service support. The latter three are addressed by Operational Excellence. In some cases, power generation and distribution and water and waste water treating are examples the market may be a monopoly. Value within enterprises in this category is much more related to bottom line performance determined by mission compliance and effectiveness, reliability, profitability, and customer satisfaction rather than top line revenue growth.

There is another value difference applicable to the implementation of Operational Excellence. Top line, revenue growth is largely a function of product development, market demand, competitive climate, marketing and sales effectiveness, quality, and pricing. All are largely strategic, and all but quality are outside the control of operating personnel. In contrast, within enterprises driven by bottom line profitability and/or mission effectiveness, virtually all of the levers of control and improvement, efficiency/effectiveness, availability, reliability, throughput, quality, cost control, etc., reside at operating levels.

There is a third consideration. Concentrating on efficiency/effectiveness and bottom line profitability produces solid results in good and bad economic conditions. When times are good, cash flow produces resources that can be invested in strategic initiatives including new product/mission definition, development, and expansion. When times are bad, efficiency and effectiveness assure survival.

Although most of the concepts are applicable to all areas, this book concentrates on Operational Excellence developed for and implemented to improve effectiveness at the operating levels of an enterprise.

JOURNEY INTO THE FUTURE

Some advocate a step change into the future as a more expedient and effective methodology to gain the optimum state. This premise would indeed be optimum if only it were possible to predict future requirements. There are many examples: could the Wright Brothers have predicted the 747? During the 1950s, many were predicting the aerocar and personal helicopters as the future of commuter transportation; a few of the former were actually constructed. Where is that today? Closer to home, there were well-known computer companies in the 1980s, which bet their survival on the personal computer being a passing fad. Needless to say, they lost! There are many more examples of ideas, technology, and products that continually emerge in the market, some with technological advantages that fail for one reason or the other.

> One well-known manufacturer of word processors stated emphatically that no one would ever want to type text on a personal computer! The well-known brand (at the time) is no longer around.

To bring this thought process even closer, consider a 1950s oil refining, power, paper, or auto industry Rip van Winkle awakening in the twenty-first century. The processes and production equipment would be quite familiar. Our friend Rip would be absolutely astounded by automation, the increased effectiveness, and reduced staffing made possible by technological advances. There is no way anyone in the 1950s could have imagined automation and control technology that is not just taken for granted today but is a competitive necessity.

Predicting the future can be quite perilous, especially for top-line-oriented, growth enterprises dependent on market dynamics and mass market appeal to build shareholder value. Creativity is even more difficult to predict. Although both were more evolutionary than revolutionary, it is probably safe to say that few visualized the iPod or iPad until they arrived in the market. In fact, probably the greatest impact of the iPod was not the technological achievement but the radical change the iPod and iTunes forced on the music industry, including the demise of many established enterprises that were unable to adapt to new market realities (anyone remember Tower Records?).

For most operating enterprises, even those in mature, stable commodity markets, forecasting conditions 5 years into the future is a stretch filled with uncertainty. Thus, improvement becomes a continuous series of bridges from today "where we are," into the predictable future; "where we must be," typically 3–5 years. The span, extent of improvement or gap, depends largely on the horizon of reasonably accurate predictions. Equally important, improvement bridges must be constructed on two strong abutments, where we are to where we must be, Figure 2.2. If either abutment, the current state or future predictions are weak or in error, the bridge will fail. Initiatives to attain the future state will likely be suboptimal or perhaps totally inadequate.

The bridge metaphor holds true only for a point in time. As Operational Excellence improvement initiatives progress, current conditions change (improve). Future requirements are also likely to change due to some combination of changes in the overall economy, market/mission, raw materials, competitive and regulatory

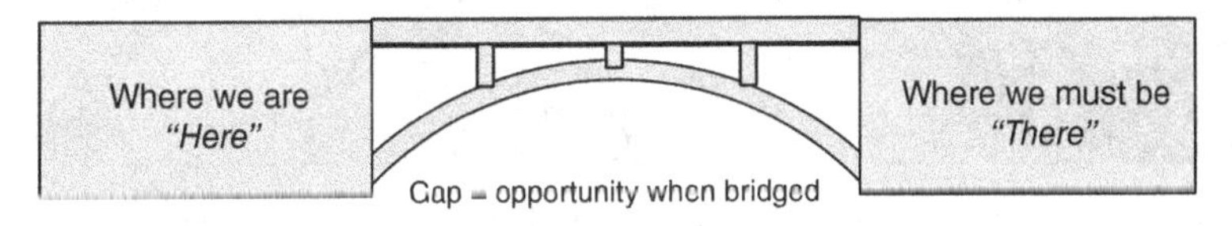

Figure 2.2 Bridging the gap between where we are and where we need to be.

environments, or a realization that the future state can be even better than thought. Operational Excellence is a constant learning process, converting learning to action. Thus, the emphasis on sustainability and continuous improvement within Operational Excellence. Strive for maximum gain, ratchet results, and continuously improve effectiveness.

Going back to aviation from the Wright Brothers to the 747, one can see how aviation advanced in steps. Each incremental step forward opened the opportunity for the next step. Answering the question of "how do we get better" at each step along the journey inevitably produced a bridge, a result that couldn't have been imagined at the beginning. Operational Excellence is exactly the same.

Although the principles are the same, every journey is different. In addition to strategic and external considerations, success requires an accurate definition and real understanding of internal conditions and how conditions may change as a function of time. The detailed assessment described in Chapter 17 is a highly recommended and very powerful method to establish internal conditions, "where we are." If, at the beginning of an improvement program "where we are" isn't totally defined and it's not quite certain where "there" might be, Operational Excellence will be implemented on guesswork, and that's not a good foundation for success or a bridge! As a final thought, Operational Excellence is a multi-year journey; the beginning and end of the bridge in Figure 2.3 is not the beginning or end of a journey. In the illustration, it is simply departing one state and entering another. Achieving Operational Excellence is a long road with many bridges. Begin with the benefits uppermost in mind and the strength of conviction to see the journey through to success!

Figure 2.3 Mike O'Callahan Pat Tillman Memorial Bridge spanning the Colorado River connecting Nevada and Arizona—completed 2010.

WHAT YOU SHOULD TAKE AWAY

Operational Excellence is not a "one size fits all." Rather, the principles must be custom tailored for the specific operating enterprise, its business/mission objectives, operating environment, position in the market, and sources of value. There is only one Toyota; don't think for a moment that what has proven so successful for Toyota will, if applied the same way, attain equivalent results for an operating enterprise in another, totally different industry segment. The overall principles are adaptable, but the relative importance, application, and sequence of implementation are probably quite different.

Think of Operational Excellence as a suit of clothes. You pick from the rack based on style, your affinity for the color, pattern, and size. From there, alterations adjust sleeve and pants length, possibly waist to fit your specific characteristics. At the end of the day, no two are exactly alike. Operational Excellence is similar; Lean Six Sigma, Total Quality Manufacturing, 5S, Continuous Improvement, Root Cause Analysis, Failure Modes and Effects Analysis, Preventive and Condition-Based Maintenance, and many other three- and four-letter acronym practices that have proven effective over many years are all part of and contributors to Operational Excellence. How and where they are applied is the key to success. This book should give you, the reader, ideas of how one or all are applied in your specific circumstances in a sequence that will *safely create greatest sustainable value* for your enterprise.

3

FOUNDATION PRINCIPLES

Operational Excellence has been defined as a management system—in reality, it is much, much more. Operational Excellence described in this book is a high performance, cooperative, success-oriented work culture. It is the way of the working life that elevates mindset, actions, and activities to safely create greatest sustainable value. As was explained in Chapter 1, Operational Excellence is an overall, master program that includes, governs, and coordinates all of an enterprise's functional improvement programs to achieve a common set of business value objectives. It relies on a set of governing principles that are constants across a broad range of operating enterprises.

OPERATIONAL EXCELLENCE—A PROGRAM EQUIVALENT TO SAFETY

In terms of the working culture, organizational, and individual commitment to program requirements, Operational Excellence is equivalent to safety. Consciously and unconsciously, everyone recognizes that safety is much more than a program, much more than a system. It begins at employment with extensive training. Safety is procedural. It avoids hazards and minimizes risk. Safety is a working culture that demands constant effort, thought, vigilance, and reminders to assure all activities are performed safely. Requirements are continually reinforced with reminders and training. Compliance is imperative.

The same is true for Operational Excellence. Operational Excellence demands focus on safe, sustainable value improvement. How can we make tomorrow better

Operational Excellence: Journey to Creating Sustainable Value, First Edition. John S. Mitchell.

than today? Conventionally, many functions within an operating organization focus on status quo; when challenges arise, problems occur; efforts are directed to restoration and correction rather than improvement. Operational Excellence demands a mindset of constant quest for improvement; identifying and eliminating cause, improving the process, practice, procedure, system, or asset, and thereby continually increasing value delivered. Challenges, problems, inefficiencies, and waste are all seen positively as opportunities for improvement. The Operational Excellence program converts the concept to actionable activities. It is the way to identify and take full advantage of opportunities for improvement to safely and sustainably increase value.

Use of Proven Practices

Over the years there have been many programs implemented addressing these same objectives. Figure 3.1 illustrates how existing functions and programs fit into the Operational Excellence matrix. Lean is primarily a manufacturing practice to improve efficiency and minimize waste. Six Sigma, Total Quality Management (TQM), and 5S began as methods to improve quality. Lean Six Sigma combines the best of both practices into one. Similar practices in the asset performance area include Reliability Availability Maintainability (RAM) analysis conducted during design, Reliability-Centered Maintenance (RCM), Preventive Maintenance (PM), Condition-Based Maintenance (CBM) or Predictive Maintenance (PdM), and Proactive Maintenance. Root Cause Analysis (RCA) is performed when the forgoing fail to provide desired results.

Most of the programs illustrated in Figure 3.1 are primarily oriented to a specific function and objective, for example, manufacturing flow, quality, and failure

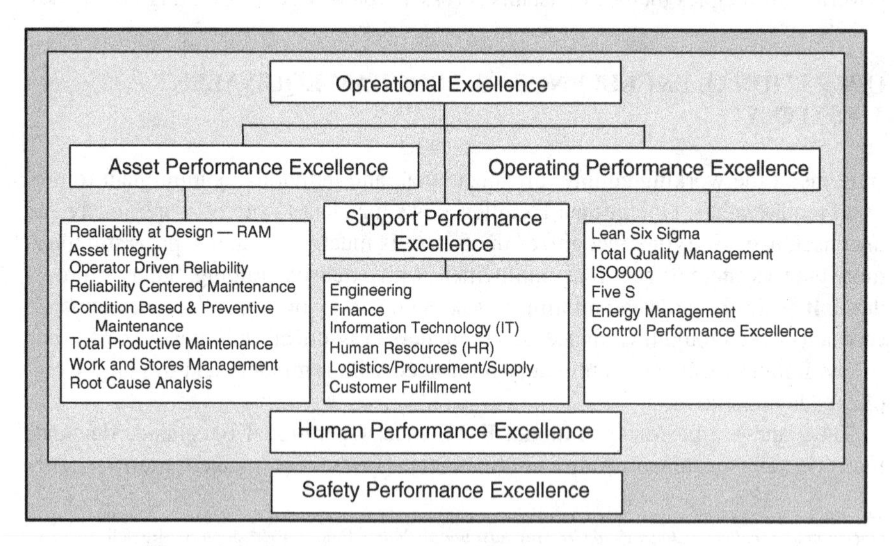

Figure 3.1 Functional divisions within Operational Excellence.

reduction. Many, Six Sigma as an example, have been expanded beyond their original application. In today's complex operating world and without a unifying concept exemplified by Operational Excellence, it is relatively easy for a functional improvement initiative created with the best of intentions to produce unintended and unexpected inefficiencies within another function or activity.

Operational Excellence doesn't replace or eliminate functional improvement initiatives but rather becomes their home. Operational Excellence is the master optimizing/improvement program for an enterprise, site, or facility. All functional improvement programs are identified, applied, and coordinated through the Operational Excellence management, administrative, and control structure described in Chapter 10. Everyone is aware of what everyone else is doing within the overall objective of safely and sustainably increasing value. Potential contributions and conflicts across organizational functions become quickly visible for discussion and cooperative resolution. Anticipated requirements within and across functional boundaries are identified early rather than becoming late breaking surprises.

SCOPE OF OPERATIONAL EXCELLENCE

The Operational Excellence Steering Team, detailed in Chapter 9, is central to achieving the cooperative working relationship that is essential to safely create greatest sustainable value. The Steering Team is composed of senior managers committed to success with the insight and power to oversee and coordinate improvement activities and eliminate the friction that inevitably exists at the interfaces within a functional organization. The Steering Team establishes overall objectives and sets the positive, supportive example for the Operational Excellence program. It approves improvement initiatives and monitors results for compliance with objectives. The Steering Team is a key ingredient to assure the success of Operational Excellence.

As an enterprise-wide program that includes every function from operations to administration, Operational Excellence can appear quite complex and even overwhelming. Similar to all complex efforts, probably beginning with constructing the pyramids, success demands dividing the overall into manageable segments. In the case of Operational Excellence, segments that will be comfortable to participants in an organization are typically divided on functional lines. With all activities governed by common objectives and structure, integrated and coordinated by the Operational Excellence Steering Team, the program can concentrate on the primary objective: using human capital most effectively to identify and implement improvement initiatives to *safely increase value*. Forming multifunction improvement action teams is the initial step of bringing diverse functions together to address and solve common opportunities for improvement.

Although all of the functional divisions will have their own champions and improvement initiatives, it is important to stress that within Operational Excellence, all operate with a unified charter and objectives, in conformance to a common administrative management and control system. All activities are coordinated by a single Steering Team. This is the only way to assure that activities are mutually reinforcing and gain the benefits of cross-function participation and cooperation.

There is also overlap and cooperation necessary between functions. As examples, Operational Excellence requires close cooperation between operations/production and maintenance. Developing and providing accurate, timely reports of gains and losses in business value require participation and cooperation from additional functions, primarily IT and Finance. Both have supporting roles in essentially all functional areas and must be heavily involved in activities requiring data and information from which to identify and value prioritize opportunities for improvement.

As illustrated in Figure 3.1, safety and human performance are vital, common elements equally applicable to all functions. This concept should be very familiar to all in an operating enterprise. It is how safety and human performance are treated in most, certainly the best, enterprises. To repeat for emphasis, Operational Excellence is the master program incorporating, coordinating, and assuring that all functions, including safety and human performance, work together effectively to create greatest sustainable value.

FINANCIAL CONSIDERATIONS

Meeting all business and mission requirements effectively and sustainably at optimum cost must be the goal of all operating enterprises. Program objectives must be aligned with and contribute demonstrable value. Illustrated in the business model contained in Chapter 5, an operating, production/manufacturing process includes numerous sources of value and associated costs. In addition to raw materials, costs include people, electricity, fuel, water, logistics, administrative and technical support, training, services such as waste removal, as well as business licenses, penalties, and fines for miscues. Other asset-intensive operating enterprises for which Operational Effectiveness is applicable will have similar value and cost elements that must be optimized for success.

Dealing with, controlling, and optimizing the lifetime cost of ownership require a clear understanding of what actual life really is and how the operating environment, operation, and maintenance affect mission and business effectiveness.

> Two concepts of safe operating lifetime cost must be given consideration. First, the overall lifetime cost from specification, through procurement, construction, commissioning, operation, and finally disposal of physical operating assets. Since operating expenses may be well over 50% of total lifetime costs, every effort must be made to optimize this major factor throughout life. It is false value to purchase physical operating assets on low initial price only to pay many times more for poor operating efficiency and unreliability. Lifetime and effectiveness between major outage/overhauls is the second factor. Improving reliability, safely extending interval between outages, and reducing the duration of outages have a significant positive impact on value gained.

Are mission effectiveness and value gain determined by mission compliance, production throughput (availability), quality/yield, market conditions (demand), customer satisfaction, operating (conversion), and maintenance costs? Typically, it will be some combination of two or more of the preceding or perhaps even unnamed others. The answer is vitally important, because it establishes the prioritization of opportunities within an Operational Excellence initiative, as well as the basis for deployment and allocation of resources. Each operating enterprise must define the exact criteria for business/mission effectiveness and value creation for their specific circumstances.

Within the organization, working-level managers, professionals, supervisors, and employees all must learn to think in terms of value and results. What are improvements (results) worth in value to the enterprise in terms of: mission compliance, production output, quality, risk reduction, cost effectiveness, and customer satisfaction? Initiatives for improvements within Operational Excellence, investments for productivity-improving technology and practices, and risk reduction must all be supported with compelling financial justification on the basis of their objectives and definition of value. Justification must be in financial terms or mission compliance that is understandable, credible, and appealing from the working level to executives and senior management. The conversion must be accomplished at the technical level. Executives and senior managers are not going to make the translation from technical advantages to business/mission success!

DRIVEN BY BUSINESS/MISSION RESULTS

Operational Excellence is a lean business/mission optimizing process. It is results, effectiveness, and value oriented rather than activity and task protective. Continuous, value-prioritized, sustainable improvement is the objective. Opportunities, prioritization, and measures of performance are linked to mission/business objectives. Safe, permanent, and sustainable increases in effectiveness, value and profits, risk, and cost reductions are gained through visionary, enlightened, improvement. Business value, profit center (management to value) rather than cost center (management to budget), principles direct the entire process. Operational Excellence incorporates and builds on the best attributes of proven practices mentioned earlier. It includes risk identification, prioritization, and mitigation through processes such as Failure Modes and Effects Analysis (FMEA) and failure and incident analysis with RCA. All are assembled and optimized to gain greatest value within specific market conditions, business and mission objectives, risk profiles, and the operating environment.

> The best enterprises recognize their deficiencies and are constantly striving to improve in every area. The rest are generally satisfied with current performance and don't see any real need to improve. "We are already doing that" is a common response of lower performing enterprises when introduced to a best practice.

All of the processes and practices necessary to assemble an effective Operational Excellence program are proven effective and readily available; however, most organizations are not realizing full value. A lack of clear business/mission objectives is one impediment to greatest success. Improvement programs may be limited to a single function and/or not effectively integrated. Poor communications and coordination between working levels where knowledge of improvement opportunities typically reside and senior management is another major reason. Professionals and people at the working level often express frustration. Many conclude that contributions they could make to corporate profitability and success are generally not considered or, worse, ignored in management decisions that focus solely on short-term cost issues. Management may think cost reductions within an operating organization can be ordered, for example, reducing number of employees. In reality, the only way to permanently and sustainably reduce cost is through increasing reliability and effectiveness that eliminate the need for spending.

> People at the working level recognize problems and inefficiencies—they have to deal with them every day. Managers who wonder why inefficiencies exist have only to look in the mirror. Most seldom ask about inefficiencies and recommended corrections. When they do, many don't listen and most don't take any real action. "Been that way since I've worked here" is a commonly heard refrain to justify inaction! When a manager inquired why no one had ever mentioned the huge waste and inefficiencies within the old system, workers replied: "You never asked!" When managers do ask about problems, listen, and initiate action taking full account of employee recommendations, the business normally runs better; people involved gain ownership and commitment to improvement.

In addition to internal mandates for cost reductions to meet profit or budget requirements, large consulting firms are often brought in to audit performance and efficiency and make recommendations for improvement. These firms typically recommend cost reductions primarily by comparison to industry benchmarks. Many do not go deeply enough into the organization to identify the causes of inefficiency or recognize that the only way to reduce excessive costs is to find and eliminate cause and improve efficiency. This leads to a fundamental truth that justifies Operational Excellence:

> It is impossible to cut costs, starve into prosperity! Arbitrarily removing costs, typically removing people, from an inefficient organization without removing the inefficiencies responsible for the costs makes the organization more inefficient and costly. Surveys confirm that less than half of companies that attempt to improve solely by downsizing fail to achieve any increase in operating profit.

Operational Excellence provides the methodology to identify and remove the cause of inefficiency. Removing the causes of inefficiency is the only way to produce permanent, sustainable improvement. Whether they call it Operational Excellence or something else, the best organizations consistently meet forecast (achieve high predictability), look for and implement improvements (better ways to do things), minimize risk, and eliminate the cause of inefficiency and waste.

> Studies consistently validate the truism that the top enterprises not only have best performance to business/mission benchmarks, but also typically have the fewest interruptions, fewest number of people, and least requirements for work.

FOUNDATION PRINCIPLES

Operational Excellence and the Operational Excellence program are formed around principles. Principles drive formulation and implementation, are mutually reinforcing, and form the basis for the working culture that is essential for success. Principles apply across varied operating enterprises and ensure the greatest possible integration of resources, effort, and contribution to mission/business value and success. All must be established and/or strengthened. The absence of, or failure to fully achieve one or more will weaken the initiative and reduce the value and benefits gained—in some cases significantly.

Principles of Operational Excellence

- Assure a safe, stable working environment
- Focus on creating maximum sustainable value
- Program strategy, objectives, activities, processes, and systems fully aligned with business/mission strategy and objectives
- An ethical, empowered working culture committed to excellence, consistency, discipline in all aspects of the culture and processes, positive attitude, ownership for success, and continuous improvement
- Effective teamwork between functions building from multifunction improvement teams
- Boldness—continually seeking new ideas, methods, and processes
- Improvement actions designed to create maximum business/mission benefits
- Optimal prioritization of all activities—minimal time spent on unimportant/low value issues
- Deficiencies and waste identified and eliminated
- Processes and procedures stabilized and improved
- Effective communications; all aware of program objectives, benefits, and successes
- Data and information accurate, secure, and accessible
- Measure what matters; results- and value-oriented key performance indicators (KPIs) rather than activity/task-based KPIs follow-up and adjust as necessary to assure compliance to objectives
- Sustain results; maintain performance measures and attention

Six D's of Operational Excellence

The eight principles can be distilled into a list of six characteristics of Operational Excellence that must be incorporated into the program to assure maximum value and success. Consider the six D's listed in the following section as the Operational Excellence equivalent of 5S. All will be addressed and expanded in greater detail in subsequent sections in this and the following chapters:

- Driven by opportunities to deliver greatest value to the enterprise—*safely and sustainably*
- Demands strong, engaged leadership—*clear vision, business/mission objectives and strategy, visible drive and enthusiasm, encouragement, and commitment to success*
- Defines program strategy and plan—*meet all SHE (Safety, Health, and Environment), business/mission objectives, legal, regulatory, and community requirements*
- Develops and sustains working culture committed to honesty, integrity, mutual trust, excellence, persistence, and continuous improvement—*organizational and individual commitment, pride, ownership, mutual support, empowerment, intolerance for deficiencies, and accountability*
- Depends on reliability—*consistency, structure, standardization, minimizing risk, and variation*
- Directed with accurate data and information—*identify and value opportunities for improvement, demonstrate results, and increased value*

EIGHT ELEMENTS OF THE OPERATIONAL EXCELLENCE PROGRAM

Principles of the Operational Excellence program are expanded into the following eight elements. Elements adhere closely to eight of the ten areas defined by ExxonMobil in Chapter 1 (the remaining two: Third-party services and Community awareness and preparedness can be implied.)

The elements are organized in general conformance with current functional organizations as the starting point for most enterprises embarking on Operational Excellence. In the scheme of things, it is far better to concentrate on necessities such as seeking value and improvement rather than beginning with less productive and potentially unsettling organizational changes. The multifunction improvement action teams chartered to identify and implement improvements will naturally bridge functional divides without unnecessarily ruffling feathers in the process.

A site manager committed to major improvement would not share his vision or plans for the future with any other than a few highly trusted subordinates. His reasoning was that the change necessary to assure continuing success within a very challenging competitive environment would be so unsettling to many employees that it could dramatically worsen morale and potentially cause collapse of the business. Changes he considered essential were made slowly and incrementally without fanfare. About a year into the improvement process, the enterprise for which the site was an operating entity came under severe profit pressure and ordered major changes including large layoffs. Because the site had already made many of the changes mandated by corporate, the additional requirements for reduction were minimal. Although not communicated, a solid vision accompanied by action produced major benefits.

Some aspects of Operational Excellence, communications and metrics as two examples, appear in multiple elements coincident with functional requirements. The remainder of this book and the assessment process introduced in Chapter 17 are organized in the same form to provide a consistent descriptive structure around which the program and improvements are identified, assembled, valued, implemented, and assessed for contribution to business/mission effectiveness.

The elements of Operational Excellence begin with the most important principle of all—results. Results are the alpha and omega of Operational Excellence, where we are (here), where we need to be (there), with the gap between the two representing the opportunity for improvement.

Results

Safely produce greatest sustainable value achieved by:

- Assured compliance to all SHE/EHS, statutory, and regulatory requirements
- Maximum sustainable contribution to top-level enterprise business/mission performance and effectiveness
- Reduced cost and risk.

Measures of value creation include the following:

- Profitability, increased throughput and quality, and customer satisfaction
- Reduced waste, scrap, failures, and cost
- Improved capital and cost effectiveness.

Leadership

Strong, engaged leadership—*leadership by example*—established and sustained at all levels of the enterprise. The importance of leadership can't be understated; positive, committed, and engaged leadership is absolutely essential for success.

- Executive-level
  - Vision—vivid, clear, compelling including business/mission objectives
  - Motivation—visible engagement and continuing drive for improving effectiveness and value
  - Continuous communication expressing necessity and ongoing support
- Operational leadership and organization
  - Committed to success
  - Assure teamwork, empowerment, ownership, cooperative working relationships—constructive work culture

- Monitor activities and results, continuing encouragement and quick elimination of barriers whenever they arise, coaching for excellence
- Minimize cross functional friction.

Requirements

Conformance to:

- SHE
- Legal, statutory, regulatory and insurance requirements, inspection, test, and calibration
- Quality (ISO9000) and other operating standards
- Operational Excellence and contributing programs organization and requirements.

Program Definition

Fully defined:

- Enabling charter, mission statement, program objectives, and strategy fully defined, published, and understood
- Organization including leadership: Steering Team, Program Leader, roles, and responsibilities fully defined; Responsible, Accountable, Supervise Consult, Inform (RASCI)
- Detailed plans directed at improving value developed by process, practice, technology, and organization
- KPIs; activity and results established
- Full commitment to continuous improvement and sustainability.

Supporting Practices and Procedures

Improve operating effectiveness, fully defined and proven:

- Enterprise/site processes and procedures
  - Risk identification, analysis, and mitigation methodology
  - Criticality (risk) categorization and prioritization
  - Failure and incident investigation analysis and avoidance (RCA)
  - Business/mission value/ROI (Return On Investment) calculation procedure
- Value, financial system
  - Capable of accurate assessment of real costs and value produced across the enterprise to the component level
- Overall improvement programs

  - ISO9000
  - Six Sigma and Lean Six Sigma
  - Five S
  - TQM and total productive maintenance (TPM)
- Functional improvement programs
  - Reliability at design (RAM)
  - Asset integrity
  - Reliability and maintenance improvement: RCM/FMEA, PM, CBM
  - Energy conservation
  - Work and stores management
  - Logistics, supply chain management.

Working Culture

Committed to excellence, continuous improvement:

- Program accepted as equivalent to safety in terms of compliance, overall and individual ownership, continuous improvement, intolerance for deviations, and sustainability
- Led at the working level by energized, empowered, engaged, and motivated champions—real ownership for success
- Fully engaged workforce—positive, trusting work environment; everyone feels valued, trusted, and demonstrates a willingness to collaborate and cooperate
- Culture of excellence in all activities and tasks, ownership and commitment to continuing improvements in every area of endeavor established and sustained
- Emphasis on improvement rather than restoration, reliability, consistency, and minimum variation
- Robust organizational improvement process
- Skills and certification requirements documented and fully supported by skills management and training processes
- Effective improvement management process
- Communications program and plan established and implemented
- Tangible rewards for results and value created.

Information Management

Complete, accurate, accessible, up-to-date and secure data, information, and documentation

- Requirements:
  - Identify current performance, performance objectives, and deviations

  - Identify opportunities for improvement
  - Document value delivered
- Full document, configuration, and revision control
- Documentation includes the following:
  - Site/facility specifications
  - Drawings, operating, and repair manuals
  - Comprehensive operating instructions
  - Current operating status, capacity, and predicted operating lifetime
  - Business/mission financial data
  - Compliance reporting.

Follow-Up

Continually monitor results, conduct periodic assessments, assure continuous improvement and sustainability.

- Conducted from an auditable formal check list
- Includes business/mission contribution metrics
- Review program mission, strategy, and performance of individual elements
- Identify additional opportunities for improvement
- Assure gains and results institutionalized and sustained.

IMPLEMENTING THE OPERATIONAL EXCELLENCE PROGRAM

Implementing details are one of the most important and often neglected aspects of any major improvement program. The Operational Excellence program described in detail in later chapters is implemented in a circular DIPICI sequence: Design, Identify, Plan, Implement, Check, Institutionalize and improve, Figure 3.2. The DIPICI sequence, described in more detail in Chapter 4, is very similar to and patterned after the Shewart Deming (Plan, Do, Check, Act; PDCA) and Six Sigma (Define, Measure, Analyze, Improve, Control; DMAIC) implementing sequences. Each step in the sequence includes specific requirements to ensure a successful program.

BENEFITS OF OPERATIONAL EXCELLENCE

Operational Excellence produces numerous benefits. These include the following:

Improved SHE/EHS Performance

Most organizations contemplating Operational Excellence have attained excellence in SHE/EHS. If they haven't, they certainly must; an organization can't expect Operational Excellence in the absence of SHE/EHS excellence. For those that have attained excellence in SHE/EHS, risk may be an additional dimension required by Operational Excellence.

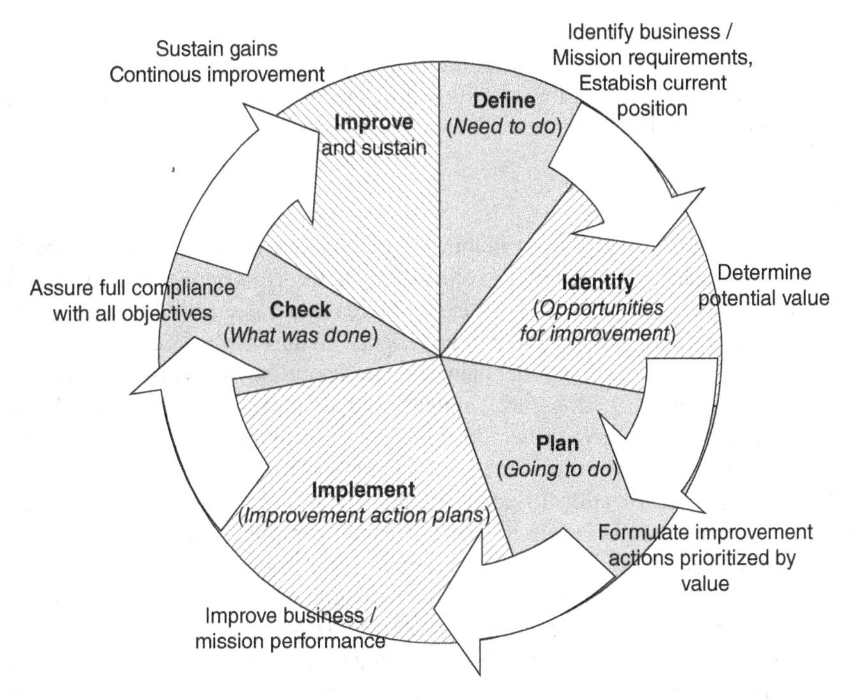

Figure 3.2 Six phases of Operational Excellence.

Reduced Risk

Operational Excellence optimizes the balance between risk, reliability, and cost and drives continuous, sustainable improvement. By doing so, Operational Excellence gains safe, permanent growth in business/mission compliance, operating/production effectiveness, and reduces spending (costs).

The probability and consequences of potential events are fully defined and considered. When a decision is being made under unusual or undefined circumstances, particularly when there is pressure to decide quickly, risk is an automatic, heavily weighted factor in the decision process. Furthermore, when a successful action is taken to mitigate an unexpected event or set of circumstances not fully covered by procedure, there is a process for post-event examination of the reaction and decision process in the context of risk to determine what additional risk might have been incurred and improve the procedure. By emphasizing risk, Operational Excellence will improve SHE/EHS performance and reduce the probability of an unexpected event that may cause an SHE/EHS event, degrade performance, and/or mission compliance.

Improved Production/Mission Operational Effectiveness

Operational Excellence identifies gaps to objective performance and develops and implements value- and risk-prioritized improvement plans that remove the cause of inefficiency. With this approach, production efficiency and effectiveness increase; incidents and failures decrease. The combination of greater throughput and compliance to mission objectives at lower costs translates directly to an enhanced mission compliance/competitive position.

Improved Reliability

Operational Excellence optimizes reliability through improvements in design, operation, and maintenance. Identifying and eliminating defects that detract from optimum effectiveness and/or require excessive expenditures in a value-prioritized sequence is a mandatory principle of Operational Excellence. Within Operational Excellence, operational and organizational deficiencies are identified for correction before adverse consequences. Actions are implemented to prevent reoccurrence. Work requirements, including low value and unnecessary activities, are safely reduced. Costs are reduced, maximum sustained return on assets achieved—permanently.

Activities within Operational Excellence that contribute to this objective include proactive, preemptive, and design improvements to operating assets and procedures, optimum operating surveillance, and high quality precision work in all functions. Operational Excellence thus provides the path to optimum reliability and permanently reduced costs, not by edict or command but rather as a result of doing things correctly. This is the only way to provide the assured availability and cost-effective business results mandated for success in today's highly competitive, cost-conscious, operating, manufacturing, and production industries.

Greater Predictability—Reduced Variation

Predictability attained through Operational Excellence minimizes unplanned, generally very expensive lost opportunity events (LOE) and "surprises," defined as any unexpected variation in mission/performance. With variation minimized, average performance approaches best performance.

Improved Capital Effectiveness

Operational Excellence provides a means to increase production/mission output and safely extends operating lifetime by reducing failures and the need for maintenance. Safe operation at greater output and longer intervals at lower costs reduces the need for capital investment in additional capacity and stocked spare parts.

WHAT YOU SHOULD TAKE AWAY

Constructed around well-defined principles, Operational Excellence is an overall, master program that incorporates, governs, and coordinates all of an enterprise's functional improvement programs. Operational Excellence provides the framework for improving process, practice, procedure, and technology around a value/risk-optimizing objective.

Customized for the specific business/mission, Operational Excellence assures enterprise objectives are gained most effectively, results sustained and continuously improved. Success requires the same organizational and individual commitment to excellence and continuous improvement, intolerance for deficiencies as safety.

Operational Excellence requires strong, effective, and committed leadership throughout the enterprise. It applies to all functions within the enterprise, with particular emphasis on operations/production and maintenance. The first to operate effectively, the second to keep the operation going, also effectively. Operational Excellence addresses the human, work culture, and organizational aspects that are crucial for success. It extends the definition and applicability of reliability and requires accurate, reliable data, and information.

Operational Excellence demands continuous follow-up and monitoring, accompanied by formal periodic reviews to assure full alignment with business/mission objectives, contribution to enterprise value, and improved effectiveness.

4

THE OPERATIONAL EXCELLENCE PROGRAM—OVERVIEW

Operational Excellence is a value- and opportunity-driven master program to fulfill the tactical and operational objectives of the enterprise business/mission strategy. Greatest success and value require establishing the Operational Excellence program as the overall governing program for all improvement initiatives throughout the enterprise. Program objectives, strategy, prioritization, and implementation are led and coordinated through a single, unified governing and administrative structure that includes an enterprise Operational Excellence Steering Team. The Steering Team is chartered to assure all improvement initiatives, including function-specific initiatives, are coordinated and mutually reinforcing to achieve greatest effectiveness and results and deliver maximum value and risk reduction toward business/mission objectives.

The Operational Excellence program transforms principles outlined in previous chapters into practice. Specifically how this is accomplished will be addressed in greater detail in later chapters.

INITIATION

An Operational Excellence program is normally initiated at the top, executive level of an enterprise. Executive level commitment and visible, continuing, enthusiastic drive are absolute essentials to create and sustain the cross-functional teamwork and cooperation that are necessary for success. An enlightened, visionary executive

Operational Excellence: Journey to Creating Sustainable Value, First Edition. John S. Mitchell.

or executives recognize the need for major improvements to assure continuing success of the enterprise and create the internal environment to make it happen. They recognize that the existing work culture, functional friction, and barriers within the organizational structure may limit the extent of progress and ultimate benefits. From these conclusions, an Operational Excellence initiative emerges, demanding broad improvements to work culture, teamwork, procedure, practice, and technology.

> Often, a senior manager will ask for and expect a "silver bullet" solution—one thing that can be accomplished to meet all objectives and resolve all deficiencies. Although not the answer expected, Operational Excellence is the silver bullet.

The Operational Excellence initiative can originate at the mid levels of an enterprise, as have so many successful functional improvement initiatives. Principles and process set forth for Operational Excellence are exceptionally powerful and can be used effectively to gain improvement objectives within a function, such as Production/Operations or Maintenance. A mid-level champion needs to make every effort to communicate and promote the principles and benefits of Operational Excellence to executive management and keep them advised of progress and successes. With demonstrated successes, the Operational Excellence functional champion has solid justification for broadening the initiative.

Cooperation between functions is the essential ingredient for the success of Operational Excellence. If Production leads the initiative, it is imperative to success that cooperation between Maintenance and Production/Operations is established at the outset so that the multifunction improvement action teams that are the heart and muscle of the Operational Excellence program can be formed and work effectively. In most cases, mutual cooperation at the level required for Operational Excellence can be established by an informal agreement between Maintenance and Production/Operations Superintendents.

Some enterprises have gone the other way, establishing essentially independent functional improvement programs in the belief that attempting to find common ground among the differing skill sets required within each function overly complicates and slows an improvement effort. This is not optimal and has been proven ineffectual. As stated in a previous chapter, objectives and efforts must be coordinated and synchronized so that an improvement program developed for one function or specific set of circumstances works cooperatively with activities within another function. A unifying structure is imperative to assure functional programs are consistent in objectives and process and work well together most effectively. Furthermore, dialog and profiting from experience allow all programs to fully use the lessons learned and solutions to challenges that have been resolved elsewhere in the organization. Although perhaps more challenging to manage, a single, fully integrated Operational Excellence program will produce consistent enterprise-wide results greater than the sum of the parts in less time and with less duplicative and wasted effort.

THE VALUE PRINCIPLE

Safely producing greatest sustainable value and meeting all business and mission requirements safely and effectively at optimum, sustainable cost must be the goal of all operating enterprises—it is the goal of Operational Excellence. To fulfill these requirements, the enterprise must define and fully understand all sources of and detractors from value. For a profit-making operating enterprise, the value process begins with revenue and cost; revenue must be greater than cost for the enterprise to survive.

> Continually seeking greatest value will eventually lead to major improvements in every nook and cranny of your organization. "Today, we have to focus on operational excellence and execution—squeezing more efficiency and profits out of the operations in an effective, prioritized sequence."
>
> Fortune 500 company CEO

The preceding section is a very simplified description. Revenue from product sales is influenced by many factors, including price, quality, response, delivery time, customer satisfaction, and of course, market forces that are typically outside the control of the operating elements within the enterprise. Reputation and good will affect revenue. In some cases (probably should be more), enterprises with reputations for prestige, quality service, rapid response, as good citizens and a loyal following can successfully sell products at premium prices so long as they meet customer expectations. Ferrari and Rolex might be premier examples!

> A product business manager was suddenly faced with offshore competition selling an equivalent product at 25% below their current selling price. After studying the situation, he concluded that close proximity to the customer, quick response, and a reputation for quality could offset approximately 8–10% of the price advantage. Accordingly, he established an objective of a 15–20% reduction in the cost of production.

This is important as it demonstrates the lengths one must go to assess the impact on value by activities and events that may not appear important to people at the working level of an enterprise. For example, an environmental incident has costs that are easily tallied; lost production, clean-up, penalties, fines, etc. What might not be so easy to tally is the impact on revenue by purchasers who decide on an alternative supplier because of loss of supply and/or bad publicity following an incident. There have been several recent events of this type that undoubtedly affected revenue of the affected enterprises.

> Value must be the key driver for prioritizing what gets done—from setting business/mission and program objectives to daily work. However, focusing only on an immediate deficiency and attendant monetary value gained by correction may hide significantly greater improvement opportunities. Thus, it is important to have a method for identifying larger, overall contributors to value. For example, a chronic problem may be due to poor operating or repair procedures. Correcting the specific

procedures will increase reliability and hence add value. Looking further, the deficiency may be more general. Although addressing the more general opportunity may not result in immediate value, it will gain long-term benefits through improved future performance. Furthermore, looking at the larger opportunity may reveal a cultural, process, or procedural deficiency that should be addressed. Action teams must be trained to look beyond specific problems to more global opportunities for improvement.

Within the value calculation, costs are generally easier to ascertain. In most cases costs equal spending; somewhere there is a record of the cost for any given activity, even if it includes multiple expenditures. Consider the environmental incident. There are clear costs for clean up, including outside contractors and company personnel that might be directly engaged. Fines and other penalties that may be applied add to the total. All can be tallied up for a total cost.

When all factors are considered within Operational Excellence, safety, health, and environment (SHE) excellence is not only required by statute, but also creates real, tangible value.

The same logic and thought process apply to nonprofit operating organizations. Although many may have a captive customer base—public transportation, water, and waste water enterprises to name three, the same principles apply. Customer dissatisfaction may affect revenue; it certainly affects costs that must be expended to service complaints and interruptions.

Thus far, value and the value principle have been stated as absolutes. In the real world, however, exceptions to the value principle may be necessary to maintain other principles of Operational Excellence, notably the continuing quest for excellence.

Two examples: firstly, routine testing results in spilling a nonhazardous material because of inadequate sample facilities. Cleanup may be relatively easy and requires 15 min or less. In a facility where the value principle rules the approval for improvement projects and work orders, the value gain from correcting this minor problem never rises above the threshold for accomplishing the work required, greatly annoying the people responsible for and doing the testing.

Secondly, a raised conveyor runs the length of a building. In order to get from one side of the conveyor to the other to perform regular duties, an operator must walk several hundred feet around the end. The solution would be to construct a walkway over the conveyor at the center. In this case, as was true in the first, the value gained by assisting an operator to move between stations does not approach the level needed to justify a simple project.

In both cases, correcting annoyances have low or minimal value when calculated in terms of business results. However, allowing an obvious deficiency/problem/annoyance to remain has tangible impact on attitude. How can people directly affected be expected to be committed to excellence and continuous improvement if annoyances that detract from the concept of excellence or efficiency are allowed to persist? Ultimately, the motivation and drive to achieve excellence are reduced.

There is a solution. Provide responsible individuals with a small, fixed annual budget to correct annoying defects that detract from consistent excellence but have minimal business value. At the end of the year, expenditures must be justified, not by a gain in business value but rather by correcting annoyances that detract from a consistent commitment to excellence. In this way, responsible individuals are empowered to identify and correct small defects with accountability for results.

From this very brief introductory summary, it should be apparent that each operating enterprise must identify and objectively determine sources of value and the value process for their specific circumstances. Not simply costs but value considering all factors including intangibles and the interrelationship between factors. This determination is the basis for Operational Excellence—how does the enterprise act to improve value and in what sequence are improvement actions implemented to have the greatest positive impact in the least amount of time? The latter to satisfy common short-term thinking and the continuing demand for "what are you doing for me today!" Operational Excellence provides the framework where short-term gains can be achieved most effectively and with lasting effect within a long-term strategy.

USE OF PROVEN PRACTICES, PROCESSES, AND TECHNOLOGY

Although it may superficially appear so, Operational Excellence does not reinvent any wheels. Operational Excellence fully uses existing, successful technology, processes, programs, and practices. Where Operational Excellence offers improvement is when they are implemented in a value-prioritized sequence to address specific opportunities. Instead of a general implementation applied to broad requirements that may not offer maximum value, Figure 4.1, technology, processes, and practices are implemented and refined to address specific opportunities, Figure 4.2. It is therefore easier to determine success. Did the technology, practice, program, or procedure create the expected increase in value? For those familiar with Six Sigma, Operational Excellence similarly identifies opportunities for improvement, prioritizes by value, then develops and implements specific action plans to gain the greatest benefits from the opportunity.

The advantages of an opportunity-driven program originating from the identification and exploitation of opportunities to *safely and sustainably increase business/mission value* and reduce risk and lost opportunity in a value-prioritized sequence are many as follows.

- Gains highest safety and environmental performance most quickly and effectively by concentrating efforts on highest priority opportunities for improvement.
- Increases business/mission value and operating effectiveness by identifying and exploiting improvement opportunities in order of value/return.
- Reduces risk through identification, management, and containment beginning with greatest risk where potential value is also greatest.
- Leads to all the processes and improvements necessary to establish and maintain best performance in a most effective, risk-/value-prioritized sequence.

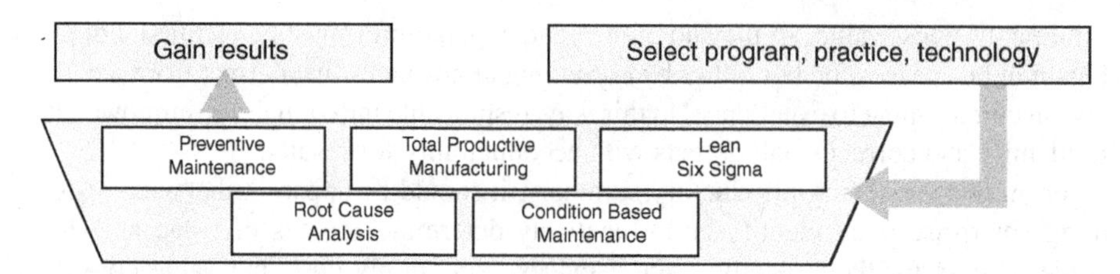

Figure 4.1 Programmatic implementation of improvement programs.

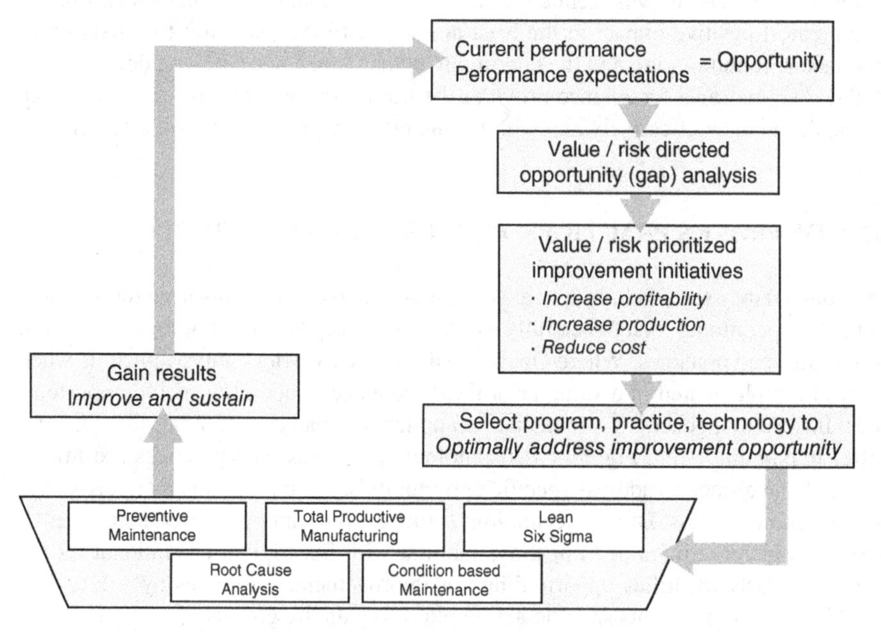

Figure 4.2 Value-driven implementation of improvement programs.

- Results in optimum operating effectiveness and reliability–minimal surprises and lost opportunity.
- Gains optimal resource effectiveness—people, material, and financial.
- Builds an effective organization and working culture—creates engagement, energy, ownership, commitment, and responsibility.
- Leads to improved practices and procedures—in an effective value return sequence.
- Demonstrates results in credible business and financial terms.
- Sustaining process.

Chapter 5 will present a business/financial model based on a for-profit operating enterprise. The financial model provides the methodology for objectively determining value and identifying all the elements of value produced by a specific improvement initiative. The principles hold true for a nonprofit operating enterprise as well.

OPERATIONAL EXCELLENCE PROGRAM

There are two basic methodologies for improving operational performance and effectiveness. The first, essentially a bottom-up strategy, begins with the premise that foundation processes must be perfected before an organization can move into more advanced, optimizing processes and practices. There is a weakness in this logic. If an operation is experiencing many problems, it is more logical to target and correct the causes of problems rather than developing robust processes to deal more efficiently with conditions that shouldn't exist in the first place.

Operational Excellence begins from a second perspective. It starts from where you are—not from where you think you might be! It is opportunity driven. Referring back to Figure 4.2, the operation is viewed from an overall perspective to identify opportunities for improvement, the top block in the figure. Benchmarks are a readily available source to utilize as targets for excellence. Deviations, gaps, between benchmark and actual performance are opportunities for improvement. Challenge is that benchmarks alone don't identify specifics for improvement. A better methodology is to use own performance. Every operation has good days when everything is going well and bad days when nothing appears to go well. What are the specific differences between the two, and more important, what are the causes? The differences can be discovered with processes such as Pareto and Weibull analysis to be explained in more detail in Chapters 12 and 13. Differences are opportunities for improvement.

> Using a Pareto analysis, one facility found that about 80% of their operating and cost deficiencies within one defined group of operating assets was caused by only 10% of the total. By simply focusing on this relatively small portion of a much larger group, improving performance to the group average, the company could eliminate several million dollars of unnecessary spending. Next step was to determine and correct specific causes. Since the focus could now be on a small portion of the whole, ranked by potential value, the task immediately became manageable.

These observations lead to another very important principle: every enterprise, every site, and even units within a site will be unique with different strengths and opportunities for improvement. At program initiation, it is imperative to define current conditions, the starting point of the Operational Excellence program, as accurately as possible. One of the best ways to accomplish this vital definition is to use the assessment process explained in detail in Chapter 17. Forming a team, likely the Program Leadership Team as described in Chapter 7, conducting detailed training in the assessment process followed by assessments themselves is an excellent way to introduce the program, ground key participants in the principles, and identify potential opportunities for improvement.

IMPROVEMENT PROCESSES

The Operational Excellence program is divided into six specific phases and executed in a progression that is identical in concept and familiar to adherents of well-known and proven Six Sigma (DMAIC), Deming Shewart (PDCA), and Boyd (OODA) processes summarized as follows.

Six Sigma–DMAIC

- *Define*—improvement projects based on business objectives, customer needs, and feedback, critical to quality (CTQ) characteristics.
- *Measure*—key processes, current performance, and defects within the processes.
- *Analyze*—deficiencies (opportunities for improvement) defined by history, risk, and lost opportunity–gaps between current and objective performance; establish why defects are generated.
- *Improve*—the process to stay within required ranges, implement optimized practices, and technology, prioritized by value and risk criteria.
- *Control*—monitor, improve, and sustain results to assure variables remain within acceptable ranges.

Shewhart (Deming)–PDCA

- *Plan*—what to accomplish.
- *Do*—initiate strategy and plan.
- *Check*—evaluate results.
- *Act*—on what has been accomplished.

Boyd–OODA

- *Observe*—opportunities.
- *Orient*—position to take maximum advantage of opportunities.
- *Do*—initiate action.
- *Adjust*—make corrections if required.

DuPont, a recognized, highly acclaimed global company has adopted a similar procedure for Operational Excellence as follows:

- *Assess*—analyze and assess the current situation, including business strategy, processes, resources, and assets. Define the future vision of what operational excellence can deliver.
- *Design*—develop a clear route map showing the actions and milestones for operational excellence, including both short- and long-term actions.
- *Optimize*—select the methods, tools, and techniques to be deployed.

- *Deploy*—apply the tools to gain improvements sequentially or simultaneously as required.
- *Sustain*—establish the fundamental process and cultural changes necessary to sustain excellent performance. Focus on performance measurement, continuous improvement, and the use of appropriate IT support tools.

THE OPERATIONAL EXCELLENCE DIPICI PROCESS

Operational Excellence borrows heavily from the preceding methodologies and adds detail essential for success. The Operational Excellence program sequence described in this book is implemented in the *DIPICI* sequence Figure 4.3 with the following expanded description:

- *Define*—the program including business/mission and program objectives, organizational and administrative requirements, and methodology for vital procedures such as risk analysis.

Note: Operational Excellence is a dynamic program, constantly evolving and improving as conditions, both internal and external, change. Thus, following program initial implementation, the Define stage transforms to review and refine at

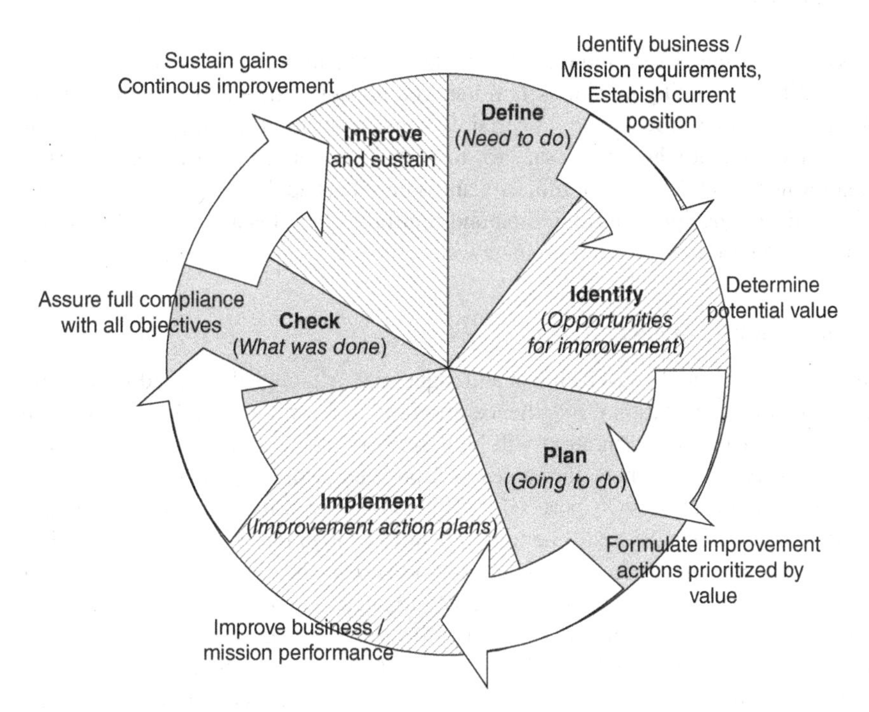

Figure 4.3 Six phases of Operational Excellence.

regular intervals. In the review process, all elements of the program including program objectives, alignment with business/mission objectives and strategy, organizing assumptions, leadership, administrative, control and management systems, and functionality are challenged and improved as required. In this way, the Operational Excellence program is itself continuously improving, remaining "evergreen" in an ever-changing business and operating environment.

- *Identify*—opportunities for improvement; prioritize potential improvements by value gain and risk reduction in two steps:
  - Analyze—identify improvement opportunities by comparison to best actual performance, performance benchmarks.
  - Prioritize—potential improvements based on value opportunity, cost, time to implement, and probability of success.
- *Plan*—improvements to create maximum value; develop detailed action plans in value order including resources and time required, responsibility and expected results (value), and probability of success; gain approval.
- *Implement*—improvement action plans, deploy resources, and technology.
- *Check*—measure and manage results; review and adjust actions as required to meet objectives.
- *Improve*—adjust initiatives as required; initiate continuous improvement, institutionalize to sustain gains and success.

The preceding is for guidance. Principles are the important elements, terminology much less so. Organizations familiar, comfortable, and satisfied with DMAIC or another similar process should certainly consider using the familiar, adding principles and content where necessary to strengthen processes that are working rather than throwing out hard won familiarity and gains. This applies throughout the Operational Excellence program; build on and reinforce strengths and use the principles of Operational Excellence to eliminate weaknesses and barriers to success.

Implementing Sequence

Descriptions in a book must be linear and sequential. Although the preceding suggests the same for an Operational Excellence program, it doesn't have to be and probably won't. A matrix implementation will be described later in this chapter. In addition to the matrix, improvement initiatives will proceed at different paces. Some, with readily apparent actions may proceed quickly through Identify, Plan, and on to Implement, whereas others may still be in the Identify or Plan phases. Furthermore, since Operational Excellence is a continuous improvement process, new initiatives will be identified and implemented continuously to augment and further gains. Thus, an ongoing Operational Excellence program will consist of a family of improvement initiatives at varying stages of maturity in the DIPICI sequence. This should be viewed as a real strength in that learning and results will facilitate and strengthen all that follow.

The DIPICI implementing sequence is described in great detail beginning in Chapter 9.

PROGRAM ELEMENTS

At program commencement, the operating enterprise identifies the business/mission and program improvement strategies, future state (objectives), overall improvement opportunities, specific strengths to build on, weaknesses, and barriers to success; the latter to include organizational, process, procedural, and even human. As the Operational Excellence program progresses, action teams are formed, and specific opportunities for improvement are identified that offer greatest value and highest probability of success. Detailed improvement action plans are developed and implemented. Results are measured continuously from implementation to completion, corrections implemented as necessary to achieve objective results; success is institutionalized and sustained.

ESSENTIALS FOR SUCCESS

The Operational Excellence program requires a number of elements to assure the success summarized in Figure 4.4. The following expands program essentials presented in the previous chapter. There is some duplication for emphasis. All will be detailed in later chapters. The sequence, and definitions within the sequence, is the basis for the assessment process described in Chapter 17.

Results

The Operational Excellence program is directed to safely produce greatest results and sustainable value.

Value gain is achieved by assured compliance to all SHE/EHS, statutory, regulatory, and insurance requirements. It mandates safe, sustainable contribution to top-level enterprise business/mission performance, and effectiveness and reduced risk. Demonstrable measures of value creation include profitability, increased throughput, minimal deviations from plan, capital and cost effectiveness for the

❑ Motivating vision	❑ Proven practices & procedures fully utilized
❑ Real, continuing executive commitment	❑ Positive, constructive working culture
❑ Clearly defined mission, objectives	❑ Engaged workforce; ownership for success
❑ Driven by value, commitment to excellence	❑ Accurate data and information management
❑ Totally committed operating leadership	❑ Effective skills management & training
❑ Fully documented improvement program	❑ Continuing monitoring and follow up
❑ Defined requirements, roles, responsibilities	❑ Commitment to continuous improvement

Figure 4.4 Elements required to assure success.

specific enterprise, mission, and operating environment. Value gained by results and improved effectiveness created at the program level link to the top-level business/mission measures through layered metrics, Chapter 15 and an enterprise business model, Chapter 5.

Leadership

Operational Excellence demands strong, engaged leadership—*leadership by example*—established and sustained at all levels of the enterprise.

Executive leaders articulate a vivid, clear, and compelling vision of where the enterprise needs to go in terms of performance and effectiveness to assure continuing success and job security. They establish the necessity for and benefits of improvement, set clear objectives and priorities for business/mission effectiveness on the basis of the vision, and assure the organization is aligned to achieve the objectives.

Executive vision must be combined with visible, continuing drive and enthusiasm, commitment to success, encouragement, and removal of barriers. All are reinforced with communications expressing the necessity for optimal execution, the importance of each person's contribution, affirming support, publicizing success, benefits to business/mission objectives, and relieving concerns. Communications are vital toward assuring acceptance, committed ownership, and continuing success throughout the enterprise. Communications and the communications plan are discussed in detail in Chapter 10.

Mid- and working-level managers must be equally committed to the success of the Operational Excellence program through teamwork, empowerment, ownership, monitoring activities and results, continuing encouragement, and quick elimination of barriers whenever they arise.

To assure continuity, any and all replacements within the Operational Excellence program leadership and management structure must be fully committed to the program and its success.

Requirements

Governing requirements begin with legal, statutory, regulatory, and insurance requirements; SHE; applicable international and national standards; local regulations; and inspection, test, and calibration requirements. Most will be required by national and local governmental bodies and insurance carriers. The enterprise must be in compliance with quality and organizational standards such as ISO9000, ISO55000, and others as applicable. All requirements must be identified by a comprehensive, auditable listing in the Operational Excellence program governing document.

Program requirements include conformance and contribution to the enterprise business/mission objectives and strategy.

Program Definition

The Operational Excellence program is defined and established with an enabling charter and detailed business plan. The charter begins with a clear mission statement and measurable program objectives derived directly from business/mission objectives. The charter and plan are periodically reviewed and updated.

The program plan continues with the strategy, specific objectives, program leadership, and organization. Program leadership and the organization/management structure include the Steering Team, Program Leader, and Operating Leadership Teams. All are defined in detail with a comprehensive responsible, accountable, support, consult, inform (RASCI) chart. RASCI charts are illustrated in Chapters 7 and 9.

> A workshop participant stated that their RASCIs were primarily a paper exercise and used as an excuse rather than a guideline. Another person commented that one RASCI is 380 lines deep.
>
> A third individual in the same workshop with a terrific sense of humor commented that their company "had a clock but no compass!"

The plan includes detailed initiatives directed at improving process, practice, technology, and organization. Improvement initiatives are identified and prioritized by potential increase in effectiveness, contribution to mission/business results on the basis of value and risk. Each plan begins with a description of the results expected, strengths to build from, potential barriers, and challenges. The plan continues with all tasks and activities required to achieve the objective; resources, risk and time requirements, specific responsibilities, and metrics/key performance indicators (KPIs) linked to enterprise mission/business objectives. A sustainability (control) plan completes the improvement action plan. Results of all improvement plans should total approximately 120% of overall program objective weighted by probabilities of success to assure meeting program objectives.

Success and effectiveness of the Operational Excellence program require a set of layered, linked metrics, Figure 4.5. Metrics create the sight line and demonstrate value creation at every level of the program. At the site level, all program metrics must link to site overall business and operational objective metrics reported to the enterprise. This will demonstrate conclusively the program value contribution to vital operating parameters including SHE excellence.

At the program level, interim and final metrics are established for each improvement initiative. Metrics include interim and final result objectives, including time lines, to measure progress and assure compliance to objectives.

KPIs are measured periodically and displayed prominently to communicate compliance to interim and final objectives, publicize success, and demonstrate value produced.

Practices and Procedures

Within Operational Excellence, proven processes, programs, practices, and procedures are implemented in a sequence and extent prioritized by value and risk. They are applied to strengthen a specific weakness, solve identified problems, and gain specific value. This is somewhat different from the conventional implementation in

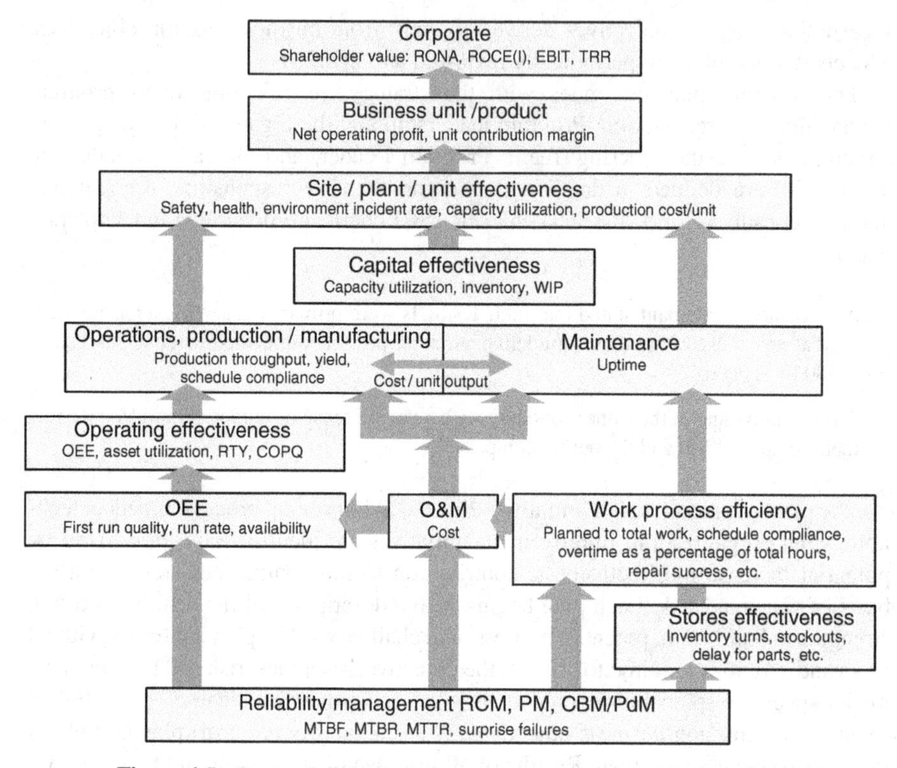

Figure 4.5 Layered, linked metrics are essential to demonstrate value.

which a program or practice is broadly implemented assuming it will create value along the way; refer to Figure 4.2.

Several enterprise-wide practices are essential requirements for Operational Excellence. These include the following:

Risk Identification, Analysis, and Mitigation Methodology Many recent events have vividly demonstrated the essential nature of a thorough risk assessment in every process, practice, and decision. Risk awareness, identification, control, and mitigation must be a guiding principle for all decisions throughout the enterprise, particularly when SHE/EHS issues are involved. An enterprise risk identification, analysis, and mitigation methodology is essential for Operational Excellence.

Tools are available to help understand and manage risks to which the enterprise may be exposed. One example is an asset criticality assessment procedure/tool to assure the risk potential of physical systems and assets is fully understood, as well as the level of effort necessary to mitigate/control the risk. External subject matter experts and programs such as Failure Modes and Effects Analysis should be used to the fullest extent to minimize risk.

Determining potential value gain likewise demands a simultaneous and thorough risk assessment. All alterations intended as an "improvement" or modification to reduce costs must include a risk assessment to assure potentially unsafe, adverse, or costly unintended consequences are recognized and controlled. Operating decisions, particularly those made under unusual or abnormal circumstances must include an evaluation of additional risk incurred. Chapter 12 contains amplifying details.

Criticality (Risk) Categorization and Prioritization The risk identification procedure described previously must be complemented by an application procedure either as a part or separate. The application procedure details requirements for classifying systems, assets, instrumentation, and work requirements by risk.

Failure and Incident Investigation Analysis and Avoidance (RCA) Regardless of best intentions and effort, no program is ever perfect. Thus, there must be provisions for identifying deviations from expectations, a procedure for reporting and documenting incidents, failures, and near misses, as well as action requirements to eliminate repetition. root cause analysis (RCA) is a key requirement of Operational Excellence. The RCA process must include the following:

- fully documented procedure for RCA;
- when applied and how applied;
- detailed methodology;
- requirements for recording, reporting, follow-up, communication, and eliminating repetition.

Business/Mission Value Calculation Procedure The Operational Excellence program must have an approved business/financial/mission value process for prioritization and valuing improvements, Chapter 5.

Functional Improvement Programs Examples of areas necessitating improvement and specific improvement programs that contribute to Operational Excellence include the following:

- Energy conservation
- Logistics, supply chain management
- Asset integrity
- Six Sigma and Lean Six Sigma
- Five S
- Total Quality Management (TQM) and Total Productive Maintenance (TPM)
- Reliability at design: Reliability Availability Maintainability (RAM); note: must include the key element of operability
- Reliability and maintenance improvement: RCM/FMEA, PM, CBM

- Work and stores management
- Organizational improvement (change management).

Practices and procedures are established in conformance with necessity, requirements, and best practices. All will include a charter document containing description, application, objectives, a flow diagram, roles and responsibilities, and metrics/KPIs. Provisions should be included for periodic effectiveness reviews. Auditable records of compliance are required.

Working Culture

Operational Excellence demands a working culture with commitment and ownership for excellence and continuing improvement. The Operational Excellence program must be accepted as equivalent to safety in terms of compliance, overall and individual ownership, intolerance for deviations, and sustainability. This requires a total understanding of the necessity and benefits for optimizing mission/business performance and effectiveness throughout the enterprise. It demands an enterprise-wide culture of excellence in all activities and tasks, and a commitment to continuing improvements in every area of endeavor that is established and sustained.

To gain the full ownership and commitment essential for success, the people who will be responsible for implementing improvement initiatives must identify and value specific opportunities for improvement, develop, implement, own and be accountable for the fulfillment plan and results. Real commitment, ownership, and responsibility require empowerment at all levels of the organization. Decision authority is based on risk; decisions should be delegated to the lowest level possible.

> Ownership and commitment are essential to successful improvement and takes time and effort to gain. Employees are typically comfortable in the status quo. They may be slow to accept management-directed changes because experience indicates the change will be forgotten in 6 months or the change itself will be changed. Employee involvement and ownership creation must be an early and essential part of the overall improvement strategy. Employees must know and believe why they must change and improve. Clear statements of necessity, strategy, and benefits are essential to pave the way toward successful improvement.

Gaining the essential working culture requires actions in several specific areas as follows.

Emphasis on Reliability There must be an emphasis on reliability, consistency, and minimum variation throughout the enterprise. As has been stated and emphasized, reliability within Operational Excellence has a far broader definition than is conventional. Reliability, consistency, and minimal variation extend to every function, every activity including the following:

- Organization, personnel, processes, procedures, data

- Design in all new construction; reliability modeling
- Asset Integrity (expanded within Operational Excellence to include all requirements for assured safely meeting all operating requirements)
- Reliability improvement programs; RCM, FMEA, PM, and CBM
- Data and information, accurate, up-to-date and secure
- Failure analysis and prevention programs; RCA.

Increasing reliability, reducing variation continually seeks to identify and improve suboptimal performance. As will be explained in detail in Chapter 13, using best actual performance as an objective, whether in an overall characteristic such as production throughput/mission compliance or within a specific population, has significant benefits and is the most effective means to drive improvement. It focuses efforts where they have most impact, will gain the greatest enthusiasm and ownership for improvement and create highest value.

> I used to be called at home every night. In the last year and a half there have been zero calls! **That's what reliability means to me!**
>
> Maintenance Manager

Robust Organizational Improvement Process The working culture of Operational Excellence demands positive human values, attitudes, and relationships. It incorporates organizational and administrative functions that have proven essential for sustainable success. Changes are generally required within a work culture to gain full acceptance for vital concepts such as continuous improvement, initiative, ownership, and accountability demanded by Operational Excellence.

The program includes an extensive improvement management process with a detailed communications plan and continuing communications to build and sustain enthusiasm, ownership, and support for the program; see Chapter 10.

Skills Management and Training Creating the Operational Excellence work culture requires skills management and targeted training. Skills management assures availability of skills, proficiency, and certifications required to implement and sustain the program. Skills must be supported with ample numbers; time and funding to assure improvement initiatives are identified and implemented effectively. A skills matrix connecting specific people to skills requirements is essential.

Training is conducted in a highly focused manner into specific areas requiring knowledge. Training ranges from a summary overview to acquaint executives and managers with program function and objectives to very specific, detailed training for people who will be directly involved with the program implementation and/or a specific procedure. After completion of training, there must be continuing opportunities to exercise skills to assure proficiency.

Tangible Rewards for Contribution Maintaining the positive, ownership-oriented working culture of Operational Excellence requires reinforcement. Many organizations praise and reward people for heroically restoring operation following problems that shouldn't have existed in the first place. People working to improve effectiveness and eliminate problems are often not recognized or rewarded.

> One facility recognized a dichotomy between principle, contribution, and reward. Rewards were typically given for heroically correcting problems that shouldn't have occurred and were often contrary to organizational principles, which stressed the importance of proactive prevention over reactive restoration. Those who worked diligently behind the scenes to prevent problems were seldom recognized and never rewarded. The facility took active steps to identify and reward individuals who were responsible for successful improvements; identifying and eliminating potential problems.

Operational Excellence demands a compensation and reward system on the basis of performance and contribution to business/mission and program objectives.

Information Management

The Operational Excellence program, individual improvement initiatives, and value prioritization are dependent on complete, accurate, secure, accessible, and easily understood data and information. Information, information management, and document control systems are vitally important to success. Information systems must be fully integrated and capable of reporting the effectiveness of operating, management, and logistics processes, as well as identifying the occurrence and potential value of correcting business/mission performance gaps, Chapter 12.

A revision control system is required to assure documentation is accurate and up-to-date.

Essential information includes site/facility specifications, drawings, operating and repair manuals, and operating instructions. It also includes current operating status, capacity and predicted operating lifetime, business/mission financial data, and the information structure necessary for compliance reporting.

Follow-Up

Continually monitoring results, periodic assessments, assuring continuous improvement, and sustainability is an ongoing process within Operational Excellence.

Performance Assessments A performance assessment described in Chapter 17 is a powerful method to establish initial conditions and performance at program initiation as part of the improvement opportunity identification process. Applicable elements are repeated as improvement programs are implemented to assess interim progress and assure improvements meet objectives.

Comprehensive performance assessments are conducted at regular periodic intervals to validate and document performance, identify additional opportunities for improvement, and assure gains are being sustained. For consistency, making what are often subjective judgments as objective as possible, performance assessments require a detailed check list or "scorecard." A scorecard extract is shown in Figure 17.2.

In addition to periodic reviews to assure progress and results to objectives, the process includes reviews of the program mission, strategy, and performance of individual elements. These mission and strategy reviews assure full alignment with the business/mission vision, objectives, and strategy as well as maximum contribution from program performance, effectiveness, and increased value.

Sustainability and Continuous Improvement Sustainability and continuous improvement are mandatory elements of the Operational Excellence program that must be built-in from the beginning. Operational Excellence improvement action plans include continuing activities and responsibility needed to build on and sustain gains similar to the Control Plan within Six Sigma (C in DMAIC). Improvement action teams may be reduced in size as gains are successfully realized; however, there must be continuing oversight, immediate corrective action if results begin to backslide. Continuing and sustaining the gains achieved must be assured. Pressures to "declare victory" and disband before sustainability is assured must be resisted. Many programs fail because once most gains are made and the objective is in sight, there is immense pressure to reduce or eliminate activities and resources. Gains and results are never sustained.

Detailed requirements and criteria for excellence are discussed in Chapter 17 and listed in the Assessment Check List.

IMPLEMENTATION

Operational Excellence is relatively easy to understand in concept and objective. But then, it gets down to implementation. There are numerous moving parts that must be formulated and coordinated effectively. The Operational Excellence program is implemented in a combination linear and matrix hybrid, Figure 4.6. Some stages are by necessity linear; others can follow multiple paths.

Once executive commitment has been established, program implementation begins by assembling a high level Steering Team. The Steering Team will define program business, organizational, and operational objectives. Most important, the Steering Team selects a program champion who will have operating responsibility for the program, the process to reach objectives and results. A Program Leadership Team is formed to refine the program and objectives. Later in the process, multidiscipline Improvement Action Teams are formed from the Program Leadership Team to identify and value prioritize specific opportunities and construct detailed improvement action plans. Plans include results expected, probability of success, time required, and cost. Having defined the action plans, the Action Teams proceed to implementation in defined steps. The results of each improvement initiative are measured and the plans adjusted if required to ensure objectives are achieved. Finally, value improvements and success are institutionalized and sustained.

Designate Executive Champion, Program Sponsor fully committed to program and program success					Define Organize
Executive Champion communicates vivid vision, necessity for improvement, benefits					
Executive Champion develops enterprise business / mission objectives, overall strategy					
Site leadership, Steering team, Program Champion/Leader, Communications Lead appointed					
Steering Team translates business/mission objectives and strategy into program mission, objectives, charter and strategy					
Steering team formulates/Validates business value, risk identification and ranking processes					
Steering Team and program leader appoint program leadership team(s), Communications team					
Conduct alignment workshop(s)					
Program Leadership Team(s) refine program mission, objectives, charter and strategy					
Communication team develops communication strategy, initiates program to develop favorable consensus and support					
Formulate Operating Plan					
Roll out Operational Excellence Program					
Program leadership team validates scope of improvements, conducts comprehensive assessment to identify areas of improvement					Identify Analyze Prioritize
Areas of improvement opportunity:	SHE	business/mission	system, material	risk reduction	
Identify potential improvements to:	work culture/org	systems/equipment	process/procedure	technology	
Identify specific opportunities by value:	risk	Pareto	Weibull	variation	
Establish current conditions to build from:	strengths	weaknesses	barriers	challenges	
Develop improvement objectives from:	best performance	comparative benchmarks		percent improvement	
List potential improvenments prioritized by:	value gain	time required	cost	sucess probability	
Define optimal implementation	step change		continuous improvement		
Identify implementing requirements	Lean six sigma	process/controls opt.	reliability, PM/CBM	work/stores mgmt.	
Expand leadership team; assure subject matter experts to develop detailed plans for each high value improvement opportunity					
Construct preliminary improvement value/risk priority list					
Appoint improvement action teams to develop detailed plans for individual and groups of improvement initiatives					Plan
conduct alignment workshop(s)					
Refine, validate, prioritize improvements by:	value	time to gain results	cost	probability of success	
Select highest value improvement opportunities for detailed planning					
Develop detailed improvement action plans	objectives	action steps	resources required	probability of success	
Establish measures of performance and KPI's with time lines connected to business/mission					
Formulate transition plan--how to get from the present to the future					
Develop continuous improvement and sustaining plans					
Establish investment/resources required	capital	expense	people	technology/equipment	
Determine extent of implementation	enterprise	site	facility/unit	pilot	
Submit detailed improvement action plans for approval					
Implement action plan requirements	capital equipment	organization	programs/practices	technology	Implement
Refine/optimize organization/working culture	excellence	commitment	ownership	partnership	
Conduct team, program and skills training					
Deploy action plan processes, practieces, technology; sequence for quick results					
Establish oversight; monitor progress and results					
Assure performance to objectives, quality	monitor KPI's	conduct training	record data	report progress	
Drive the implementing process; overcome barriers					
Continue training; improve proficiency	team	program	process	skill	
Communicate activities, promote results	sucesses/wins	people contribution	lessons learned	expectations	
Measure and monitor results	assure progress	meet objectives	validate program	identify opportunities	Check
Confirm contriubution to program, enterprise business/mission objectives					
Intiate continuous improvement	seek opportunities	refine, adjust, extend actions		expand ownership	Improve Sustain
Sustain gains	initaiate sustaining plan	monitor KPI's	prevent slippage	maintain ownership	
Overcome resistance	training	counseling	peer pressure	assign other duties	
Review and refine assumptions	business / mission	program objectives	value gain	support and ownership	
Results: Safely produce maximum sustainable business/mission value					

Figure 4.6 Matrix implementation.

As illustrated in Figure 4.6, the initial, Define phase follows a linear progression where one step, such as appointing the Steering Team, leads directly to the next. As soon as the Operational Excellence program reaches the Identify stage, it branches into a matrix of choices depending on the specific opportunities and the necessities for exploiting the opportunities. As an example, the processes for organizational or procedural improvements will be quite different from those to achieve material improvements.

The next, Plan phase, will be a combination of linear and matrix choices, again depending on the specifics. The remainder of the program will be implemented in a matrix depending on the specific action plan, technology, processes and resources deployed, and results expected.

The combination linear and matrix implementation is an exceptionally valuable characteristic within Operational Excellence that differs considerably from a typical programmatic implementation. In a typical implementation, an improvement program such as RCM is selected for assumed overall benefits. People are trained and the program broadly implemented without much thought to value prioritization (although this is changing). In practice, if the people trained don't immediately connect the program to solving their specific challenges or aren't able to use the program within a reasonable time following training, enthusiasm and expertise decline rapidly with a resulting waste in time and resources, as well as suboptimal results from the program itself.

In Operational Excellence, a supporting program, practice, or technology such as RCM is selected to address a specific opportunity for improvement. The team identifying the necessity for RCM is trained and immediately puts the training to use addressing their specific opportunity. This concept of opportunistic implementation is far more effective to gain enthusiasm and ownership for the process, demonstrate results, and achieve maximum value as early in a program as possible. People are engaged working on real opportunities for improvement they have identified and for which they are enthusiastically committed to attain. Practices and technology are implemented to address specific deficiencies rather than applied broadly. Results can be seen quickly. The expertise gained can then be distributed within the Operational Excellence program as the need arises. Assuring that processes and advances implemented by one initiative to take advantage of specific opportunities are effectively transferred to other areas as opportunities are discovered is a challenge that the champion must be capable of managing.

A major improvement program at a large site based on a generalized process without effective prioritization did not achieve sought after results. The most experienced participants, who were essential to success, concluded the program did not address problems they faced on a regular basis, was not making their life less complex or any easier, and not worth the time expended. Most rapidly dropped out. Moral of the story: to achieve success, an improvement program must address and solve real problems, capitalize on real opportunities for improvement that align with business/mission objectives **and** are important to participants.

Recognize also that as the Operational Excellence program advances, it will consist of many improvement initiatives at varying stages of maturity, each with a Figure 4.6 representation. Thus, an Operational Excellence program is really a matrix of organizational, practice, and technology improvements spread through a series of initiatives. Although the program presents a management and coordination challenge, the enthusiasm, ownership, and effectiveness create inestimable real value on a compressed time scale.

WHERE AND HOW TO BEGIN?

A long-term management commitment is essential to provide the direction and resources necessary to fully implement the Operational Excellence initiative. Unfortunately, this is not always the case. Management and/or priorities can and often do change. Although Operational Excellence is a long-term sustaining concept, continuing short-term results, "quick wins" are imperative in order to build and maintain enthusiasm, confidence, credibility, and support. As in so many activities, long-term support and funding are often dependant on the achievement of solid, short-term results.

It should go without saying that the program starts from where you are *(not where you may think you might be)* and gets you where you need to go. As stated, for best and fastest results, Operational Excellence must begin with a comprehensive assessment (described in Chapter 17) to establish initial conditions. Operational Excellence will then:

- Build on current strengths
- Utilize institutional knowledge and ownership to greatest advantage
- Identify and implement opportunities for improvement in a value-prioritized order
- Strengthen weaknesses
- Overcome barriers to success
- Produce real results and value in the shortest possible time.

A number of fundamental issues must be considered as the implementing strategy is developed. Mission, business and economic conditions, and the resulting executive strategy and priorities have a way of evolving over time. The implementation plan must have sufficient flexibility and adaptability to contribute continuing value and results to the business/mission despite changing conditions. As stated earlier, periodic reviews of the program itself, changes and continuing improvements to optimize performance are an integral part of keeping the Operational Excellence program evergreen.

When existing processes and practices are close to objective performance, a continuous improvement optimization process can begin immediately. But when vital functions, processes, and practices are absent, or there are major deficiencies compared to objective performance, a step change is required into the new processes. The step change improvement process is described in Chapter 10.

Ultimate success will be determined by how well the strategy and implementation plan addresses the key areas with the greatest impact on the following:

- Safety and the environment
- Events, conditions, or deficiencies that have an effect on business/mission/operating performance, effectiveness, and profitability

- Gaining performance objectives in business/mission compliance, availability, yield, quality, and cost effectiveness
- Ability to take maximum advantage of opportunities for increased production and/or quality
- Improvements in process effectiveness, reduced cost.

The business model presented in Chapter 5 is a starting point. An economic analysis enables improvement initiatives to be developed and prioritized in terms of real value. What is the potential value of available courses of action—including no action? How much does each improvement action cost to implement? What is the probability that improvement initiatives will achieve their objectives? What is the return in financial terms such as Return On Net Assets (RONA), Return On Capital Employed/Invested (ROCE/I), or some other enterprise measure of value gained? In virtually every case, far more opportunities for improvement will be present than there are time and resources for their implementation, hence the essential nature of value prioritization.

Considerations for Commencing with a Pilot

When the step from current conditions, processes, and/or procedures into the new is large, a reduced scale pilot is an attractive alternative to consider. A pilot can also be considered to minimize risk for any new initiative. A pilot provides the opportunity to test a new program, process, or procedure, identify and solve any unforeseen problems on a small scale, and thereby minimize risk. More details will be found in Chapter 14.

A pilot can be implemented in a site or unit within an enterprise, an operating unit or area within a site, or any other definable entity. Selection criteria should include awareness of the need and enthusiasm for improvement. Since success is an essential ingredient for continuing support and commitment, a pilot should be planted where it has the greatest possibility for success.

There is one other issue that is best accommodated with a pilot. Improvement initiatives may require changes to infrastructure, processes, and procedures. These are typically handled best in a pilot implementation where exceptions can be allowed, approved, and refined without disrupting an entire organization.

Third Party Facilitation, Assistance

An initial question that must be answered during the formation stage of an Operational Excellence program is whether to rely exclusively on internal resources to implement the program or partner with a specialized third party. A qualified third party brings expertise, the ability to move a program forward quickly with greatest efficiency and fewest missteps during the learning process. A third party must be able to demonstrate sufficient flexibility within their method of program implementation to use existing strengths and address specific weaknesses as effectively as possible.

Some factors to consider making this important decision include the following:

- Will the third party identify site/unit specific strengths and opportunities for improvement; how will this be done? An initial assessment should be considered.
- Will the primary focus of the third party be on a single aspect of Operational Excellence or the overall holistic process?
- Will the proposed program be constructed around site/unit specific opportunities for improvement and most rapid realization of results?
- Will the third party bring mentoring expertise to facilitate and train site personnel or will the third party perform the bulk of the work? The first is highly preferable to create the skills and ownership necessary to continue, expand, and sustain the Operational Excellence program following departure of the third party.

There is another element to consider: compensation of the third party. Fixed fee plus incentive for results based on an accepted, value improving, formula should be considered.

WHAT YOU SHOULD TAKE AWAY

An Operational Excellence program is most effectively initiated from the executive level of an operating enterprise where real power exists to motivate for success and clear barriers. The program can be successfully initiated at mid levels with effective communication both up and down the enterprise to publicize necessity and success and build support.

A positive work culture and cross-functional cooperation are essential for success. Developed and reinforced by multifunction improvement action teams, the Operational Excellence program contains all the ingredients to promote the essential work culture and cooperative working relationships throughout the enterprise.

The Operational Excellence program is directed to achieving maximum sustainable value and continuous improvement. Value in Operational Excellence includes risk and is defined for the specific enterprise strategy, market/mission environment.

Operational Excellence is a master program that applies and most effectively uses proven practices and technology. Practices and technology are applied to address specific improvement opportunities, thereby quickly gaining support and proficiency. The program is implemented in a familiar phased sequence. Some phases are implemented linearly, others in a multipath matrix.

The program begins from current conditions accurately determined from a formal assessment process. It delivers results in an effective value-prioritized sequence. Beginning with a program pilot should be considered to maximize learning and the opportunities for success while minimizing disruption to established procedures.

5

BUSINESS AND FINANCIAL ELEMENTS

Today's operating, process, and manufacturing environment is characterized by increasing intensity, more restrictive constraints, and huge pressures to reduce costs. All necessitate more concise, value-oriented methods to identify opportunities for improvement, prioritize and align activities, and to measure and continuously improve results. Identifying value requires a model that provides an accurate and objective means of calculation for the specific business/mission and external environment of the operating enterprise. The business/financial model introduced in this chapter is primarily designed to provide a means to calculate the value of essential methodologies and potential improvements to prioritize opportunities, justify investments, and financially credit results gained by Operational Excellence. It is designed for profit-making enterprises; the principles can be adapted for mission-oriented enterprises.

CONNECTION TO BUSINESS RESULTS

Technology and practice have developed to a level capable of recognizing most anomalies and defects in time to prevent outright failure and minimize unscheduled operating interruptions. However, measures of effectiveness have tended to remain subjective and intangible: do processes, systems, and physical assets perform as required, when needed, at design efficiency? Are unexpected failures and operating interruptions few and far between? Are problems corrected promptly; are operations/manufacturing superintendents happy?

Operational Excellence: Journey to Creating Sustainable Value, First Edition. John S. Mitchell.

Many of the measures frequently cited to justify advanced management technology, systems, and practices are largely intangible and disconnected from business and financial results. Do they improve operating effectiveness? Can the real value created be determined? More sophisticated operating enterprises have gained interdepartmental agreement for an average cost of a variety of events, including lost production. Although this adds some objectivity, it does not answer the basic question.

There is another challenge. After years of successful improvements, performance improvements will slow, and spending will inevitably flatten at a lower rate. At that point, some executives assume that most potential improvements have been harvested. Thus, the programs and activities responsible for these results can themselves be converted into value by reducing personnel and funding. Can this be true? Are gains in production effectiveness and cost reductions from enlightened programs, practices, and technology permanent, or are sustaining efforts required? This is a very difficult question to answer without historical measures of performance linked directly to enterprise effectiveness and profitability.

For all these reasons, investments for promised improvements in efficiency—better results from existing processes—can be justified. However, justifying investments for new technology and methods that add to or alter the way things are done to improve effectiveness, attain greater overall results and value may be more difficult. This is particularly true of efforts to convince executives obsessed on reducing costs. An accurate and credible financial model that demonstrates potential value gains from improvements and includes methodology to accommodate external changes such as mission compliance, market demand, and pricing is an imperative of Operational Excellence.

Performance Measures

Overall performance measures that combine operating/mission compliance, availability, production output, and lifetime cost are necessary for prioritizing resources and assessing the effectiveness of optimizing efforts. Measures must originate from business/mission requirements, objectives, and market conditions. They must point to and provide value justification for increased effectiveness and profitability and lead to optimized decisions. They must be equally applicable for an entire enterprise, operating unit, as well as individual systems and components.

Several overall measures of performance are in use. Cost as a percentage of Replacement Asset Value (RAV) is often used in benchmarking and as a performance metric in the hydrocarbon industries. However, this metric requires a consistent method of calculating RAV (defined as the current cost to duplicate production output/capacity) and does not consider operating environment, intensity, and age, all of which affect operating cost. Overall Equipment Effectiveness (OEE), from Total Productive Management/Maintenance (TPM), is another often-used measure of operating performance. It measures availability, production rate, and quality as a percentage of objectives. It does not consider the cost to sustain results, attain and sustain improvements. Omitting costs is a major shortcomingof this metric that is addressed in the financial model introduced later in this chapter.

One other observation about OEE as a singular value: a negative change in one element can be masked by a positive change in another.

> One industry leading site recommends monitoring the three elements: availability, production rate, and quality as separate quantities rather than a combined OEE.

THE OPPORTUNITY

As mentioned in the previous chapter, improvement initiatives are typically identified and implemented programmatically. That is, a specific program, practice, or technology is selected for its virtues applied to a perceived need, procured, and broadly implemented. Although this method certainly results in improvements, there are two major deficiencies. First, a typical improvement program addresses many areas where improvements may not be necessary and/or produce sufficient value. Second, many improvement programs are either technical or organizational. While value gain is implied, it isn't the primary factor. The program is seen by many as the end in itself. Implement the program and gain value rather than increase value through the program, practice, or technology. Although the difference may appear to be a minor issue of semantics, it is huge in terms of mindset, application, objectives, and even results metrics, Chapter 15.

As has been stated and will be repeated, Operational Excellence uses technology and programs that have proven effective over decades of application in a wide variety of operating environments. The value focus of Operational Excellence turns selection and implementation upside down. Opportunities to create additional value are identified and determine the selection of improvement programs, practice, and technology. Program, practice, and technology are implemented only where and to the extent that increased value can be demonstrated. Success is dictated by business/mission value—financial results. Engineering judgment is extended by the burden of financial proof. "Show me the money" has become substantially more than a catchy line from a popular movie. This vital element of Operational Excellence will be explained in greater detail in Chapter 14.

PROFIT CENTER MENTALITY

"Show me the money" leads to another concept: the advantages of shifting from a cost-centered, manage to budget constraint, to the profit-centered, manage to value mindset. To repeat for emphasis, within a profit-center, investments and added costs are evaluated from the standpoint of value and return. A profit center mentality promotes initiative, agility, optimization, ownership, and investment for improvement. This mentality will clearly produce superior results in a complex operating or production enterprise. In contrast, a cost center contains no systemic incentives to optimize or improve.

> If anything, there are institutionalized disincentives to optimize a cost center. When congratulations have gone cold, everyone recognizes the rewards for performance below

budget. The under budget is combined with the planned reduction and voila, the new budget objective!

Pressure to reduce costs orients an organization toward protecting tasks and activities rather than directing efforts to improving effectiveness and overall results.

Justifying Improvements

Many enterprises require justifying improvement initiatives with a calculated Return On Investment (ROI). ROI has proven elusive as a measure of increasing effectiveness. In many cases, records, before and after improvement, aren't sufficiently robust and accurate to determine value gained. Some who proudly report a high ROI for specific initiatives do not experience a corresponding improvement in performance in the financial statement.

An executive at one operating enterprise stated that he had a drawer full of optimistic project ROIs but had not yet observed any financial improvement at the bottom line!

A few have made a comparison with companies in the same industry they know are not spending an equivalent amount on advanced practices and technology; in many cases, bottom line results are comparable.

Why is there a difference between expectations, common measures, and bottom line results? Some reasons are as follows:

- In general, within many enterprises, there is no valid way of connecting small improvements (as opposed to large projects) to results. Did a given investment produce the expected financial results and if not, why? Most enterprises track budgeted versus actual expense for large projects. Very few have the information, tracking, and accounting structure to accurately determine value gain, including the profit/cost impact, of improvement initiatives. This is an imperative of Operational Excellence.
- The best technical professionals are passionate, often overly optimistic, and may be totally oriented to technical results. Many have little appreciation for, or even interest in, the results of their work in terms of business/mission value. In times past, optimistic expectations and subjective benefits were sufficient. If an improvement produced the expected technical results, there was little or no effort to determine overall business/mission value. If costs far exceeded value returned that was OK. Chances are that it wasn't noticed so long as technical aims were achieved. This is no longer true. As stated earlier, "Show me the money" is now the way the improvement game is scored.
- Conventional ROI calculations for improved processes, technology, and practice typically do not account for variations in market, business, and operating conditions as will be detailed later. Changes in external conditions can have a significant impact on the return gained from an improvement initiative. Operational Excellence demands a sensitivity analysis as part of the justification to

demonstrate a proposed improvement will generate a positive value and return under any reasonable forecast of external conditions.

There is a major requirement for accurate, traceable information, such as lifetime cost, Mean Time Between Failure (MTBF), or equivalents, for each individual system, asset, and even component within an operating enterprise. The exact cause of a failure, components involved, and the cost in terms of lost production, restoration to service, and any penalties involved is imperative information that must be sortable by system and asset, cause, component, manufacturer, model, and other criteria to detect site and enterprise-wide patterns. If improvements are made in operating practice or materials to name two, there must be a way to match results with expectations. If results do not agree with expectations, information must be available to determine why.

Some are suggesting that so-called big data will be able to compensate for a poorly structured operating hierarchy. In this case, as in many, the term *garbage in, garbage out applies*. Hierarchies are addressed later in this chapter and again in Chapter 9.

Value Prioritization

An effective business value model for Operational Excellence must include the ability to prioritize the application of resources by value creation in terms of business/mission objectives within an environment where opportunities far exceed resources. The model must be capable of comparing actual results to expectations, especially when improvements and results may span a considerable period. For example, the full results of improvements driven by a systematic, enterprise-wide program of root cause analysis (RCA) and defect elimination may not be seen for several years. Segregating improved members of a large population into a separate classification of improved is one method to identify results sooner.

Identifying the potential value recovery by elevating poor performers to a population average (e.g., MTBF) uses facility data. It creates enthusiasm and ownership among the people involved and is typically more credible to finance compared to citing industry benchmark performance as the objective.

Many of the concepts presented require a great deal of testing and refinement to best serve specific enterprise conditions. The central theme: that any investment for improvements in reliability, risk, organization, practice, and technology must be traceable to increased value in terms of business/mission performance is indisputable.

VALUE IMPERATIVE FOR OPERATIONAL EXCELLENCE

Definition

Operational Excellence defined in Chapter 1, requires optimum availability, yield, quality, and costs to achieve full, safe, and sustainable compliance with mission/operating requirements. Optimum is the operative word meaning greatest effectiveness. Not necessarily least cost but a balance that produces greatest business/mission value when all factors are considered. To repeat a primary message: Operational Excellence is directed to *results* rather than *activities*.

With this perspective, sights must be elevated to all factors that influence the creation of lifetime value. Arbitrary cost reductions are counterproductive if they lead to diminished value by compromising business/mission success through some combination of decreased availability, output, yield, or quality and/or increased costs.

A fundamental precept of Operational Excellence is the recognition that for most, financial results are the fundamental measures of enterprise success and the specific contribution of improvement processes and programs. Operational Excellence includes all factors that determine and influence lifetime cost of ownership. Effective design, installation, and operation are vital elements of effective Operational Excellence. A study asserts that as much as 60% of nonoperating lifetime costs are caused by preventable problems including faulty design, installation, operation, and maintenance practices. Not surprisingly, surveys consistently demonstrate that best-in-class operating enterprises gain the best OEE and return on assets as well as least costs and unscheduled downtime.

Financial Orientation

Analysts and investors state that the value of a profit-making enterprise is driven by its ability to generate increasing levels of free cash flows (FCFs) year after year. FCF is the level of cash flow generated by a business in excess of the costs and investments necessary to sustain the current activity.

A clear and credible connection between lifetime cost and profitability/mission compliance must be established within Operational Excellence. In a profit-making enterprise, capital-based metrics such as Return On Net Assets (RONA) and Return On Capital Employed/Invested (ROCE/I) are considered better measures of value creation process than ROI. Some companies use some variation on Earnings Before Income Tax (EBIT). In terms of evaluating the performance of specific operational excellence opportunities, the capital-based metrics offer a far better statement of effectiveness, contribution to mission, and business objectives, than cost-based measures such as cost as a percentage of RAV for the reasons mentioned earlier.

Within Operational Excellence, success begins with a change in mindset from reducing cost to safely gaining maximum sustainable value and profitability from production operations. The vitally important profit center versus cost center mentality mentioned earlier.

THE FINANCIAL STATEMENT

Revenue from sales, the top line of a financial income statement, is the net business value received for products produced by an operating enterprise and sold into a market. The cost of goods sold is subtracted to arrive at gross profit. Cost of goods sold includes raw materials, production labor, maintenance costs, utilities, etc. Gross profit as a percentage of revenue is a financial measure of production efficiency.

Note: For the purposes of this discussion, market price, a major factor that determines revenue, is dictated by a number of elements outside the production process. These include market saturation, product features, quality, reputation, and availability. Vital as they are to revenue, they are not considered in this discussion limited to considerations within the control of the operating enterprise. For the purposes of defining operating/production effectiveness, costs such as general and administrative (G&A), marketing, depreciation and others are not considered. The one exception might be engineering.

A decline in gross profit typically results in orders from the executive level to reduce costs. This can occur despite what the real cause may be deteriorating prices in a competitive and/or saturated market. The following is the crux of the necessity for operating/production effectiveness.

To be successful, an operating enterprise doesn't have to be lowest cost producer. The operating enterprise must be able to achieve a satisfactory gross profit at the least demand and pricing anticipated within a given market. For example, if least market demand is forecast at 70% of maximum, the fixed cost structure of the operating enterprise must produce an acceptable profit at least volume and anticipated price. In this way, higher cost competitors will be forced to absorb all or nearly all of the reduced market demand.

SELECTING FINANCIAL MEASURES OF PERFORMANCE

Financial measures of performance are essential to demonstrate the full value of Operational Excellence toward mission compliance, operational effectiveness, and profitability. Financial measures must have five primary attributes as follows:

- *Credible*—to business and financial executives who may have little or no appreciation for the potential contribution of optimizing processes, practice, and technology toward the creation of enterprise value.
- *Measureable*—from values within the enterprise business/financial system.
- *Accurate*—represent the value of increased utilization and effectiveness, taking into account mission and market opportunities for increased production and/or quality, product margins, manufacturing performance, and cost.
- *Impartial*—arbiter that indisputably demonstrates the necessity for, priority of, and enterprise profit impact of investment to improve operating effectiveness, eliminate defects.
- *Inspirational*—promote commitment, ownership, enthusiasm, and a profit-centered mentality.

Ideally, the financial measure or measures apply top to bottom within an enterprise. The measure used by a senior executive focused on shareholder value must be consistent with and linked to measures used by line management, engineers, process operators, craft, and support personnel. All must understand the strategy, priorities, their individual contribution, and how it creates value. Financial measures must provide clear direction and demonstrate the necessity to meet quality standards and perform assigned tasks effectively.

Team athletics provide a good analogy. Everyone on the team must be focused on the final score. Individual statistics, no matter how overwhelming, are of no use if the team does not win. In fact, the lucrative individual incentives offered to many highly paid professional athletes often have a negative impact if not directed toward team victory. They concentrate efforts on self rather than team. Business rules in the production and manufacturing world and profit is the score! Another trite saying: "There is no I in team!"

> A large consulting organization concluded that professional team sports offered a unique opportunity to assess the differences between winners and losers at the highest levels of proficiency. Answers from players were not unexpected. When they interviewed locker room attendants, people whose jobs were picking up after messy athletes, keeping uniforms and equipment clean, they were very surprised. Attendants for the best teams stated their primary job was keeping everything orderly so the athletes could devote full attention to winning a championship. Some of the locker room attendants had a better grasp of the real objective than the athletes!

A representative value model must depict a real relationship between reasonable expectations for improvement and the impact on overall financial/mission performance. The model that emerges from this concept must provide the ability to predict and, therefore, tune a process to achieve greatest effectiveness and value and mitigate deficiencies identified from the model.

A value model designed to account for all of the value and benefits generated by Operational Excellence will be described later in the chapter.

ACCURATE LIFETIME COST TRACKING

Essentially, every modern operating enterprise has some level of an Enterprise Resource Management (ERM) system in place and in use, providing the structure and storage for financial and Operational Excellence data and information. Full description and function of this system are well outside the scope of this book; however, several highly important points must be mentioned.

It is absolutely imperative to have the ERM system structured to accurately identify potential value gained by improvements and increased effectiveness as well as the real cost and consequences of deficiencies. The system must be able to associate and track total costs consumed by systems, individual assets, by manufacturer, specific model, and part number separately and grouped by category and type. ISO 14224 (Petroleum, petrochemical, and natural gas industries—Collection and exchange of reliability and maintenance data for equipment) recommends a nine-level taxonomy down to part number. The ability to sort cost data in a variety of ways is essential to identify deviations from required performance and provides vital input into

Pareto analyses (Chapter 12) of cost times occurrence that are necessary to prioritize improvement initiatives.

A word of caution, equipment asset, component, and part manufacturers often assign different part numbers to the same component. For accurate tracking, identifying good and bad performers, it is essential that identical assets, components, and parts can be identified.

Properly configured, the ERM system provides the vital information necessary to gain maximum effectiveness by identifying inefficient and/or failure-prone systems, assets, and components for improvement. An accurate, representative financial management system is essential for the success of Operational Excellence.

As much as 80% of the total lifetime cost of assets and equipment is expended during operation. In order to avoid low initial cost to dictate purchases of new equipment, the most enlightened enterprises use total lifetime cost to compare suppliers. The comparison includes actual service performance as well as training costs to master new equipment and systems. It is a much more accurate and representative means of awarding purchase contracts than simply low initial cost.

THE BUSINESS VALUE MODEL

The business value model is defined for an operating enterprise in which the value created between the price of delivered goods and cost of operations is a prime measure of shareholder value and successfully remaining in business. The concept is applicable to an enterprise, operating unit, facility/site, or individual units in a multiunit site. At some sites, the output from one unit is the input to another. Under these conditions, the calculation of transfer prices is all-important to ensure an accurate, representative picture of value creation for every definable operating unit.

> One organization testing the model stated that the model agreed with actual results and added a great deal of insight into exact cause for noncompliance with objectives. They liked the concept of combining OEE and cost into Overall Operating Effectiveness (OOE, explained later) and commented that results were greatly dependent on how aggressively objectives were set. Another company testing the model determined that completed and planned reliability improvement projects improved a financial measure of effectiveness by nearly 8%.

The model is designed for production and manufacturing enterprises; however, the concept and principles can be applied to operating organizations such as water treatment facilities, public transportation, and the military where cost-effective mission compliance is of prime importance.

Figure 5.1 illustrates the Business Value Model, or simply Value Model, its major elements, and their relationships. The top tier represents the business process. It is constructed as a simplified income statement that begins with revenue from sales and concludes with a calculation of capital-based value performance metrics, RONA and ROCE or FCF. Other financial metrics can be calculated. It is absolutely essential that the terminology, categories, flow, and output of the Value Model conforms exactly to financial statements used in your organization.

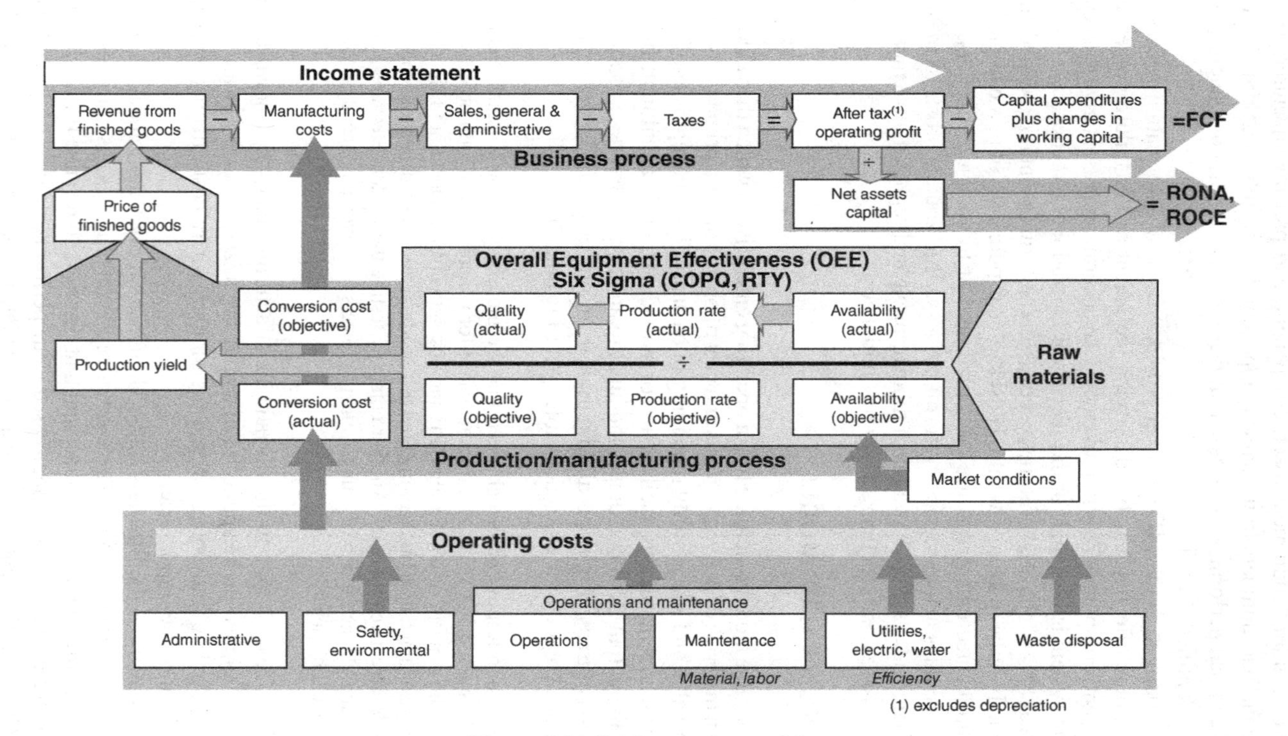

Figure 5.1 Business value model.

The middle tier represents the production process using OEE combined with conversion cost effectiveness as the prime measure of production effectiveness.

OEE is a normalized quantity representing production efficiency. It consists of three components multiplied together: availability, production rate, and quality—all expressed as a percentage of objective.

The middle, production layer, and the top, business layer are linked by production yield multiplied by the price of finished goods. The product of the two, revenue from sales, is the top line of the income statement. The Value Model uniquely links production and cost effectiveness, to revenue from sales and the business process.

As stated earlier, the price of finished goods is established by many factors outside the realm of Operational Excellence. Overall economic and market conditions and competition are just three factors that influence revenue from sales. Operational Excellence focuses on improving operating and production effectiveness so that someone else has to absorb a decline in market demand. Note also that market conditions influence the value of availability. If production output is less than capacity for any reason, availability has less value than it does when production is sold out. There is also a waste component to availability, run rate, and quality that is applicable independent of market conditions.

Within the Value Model, conversion costs are defined as "inside the fence" costs required to produce a given product. Categorized conversion costs are itemized in the lower tier of the model. These include all areas of production support costs including utilities, Operations and Maintenance (O&M), safety and environmental, administrative and waste disposal, and penalties and fines for transgressions; there are probably others depending on the specific business/mission.

All value contributed by the Operational Excellence program through streamlining processes, greater effectiveness, etc. must be captured in the model. Most supporting processes have costs and produce value. As an example, an asset reliability improvement program initiated in maintenance can improve availability, reduce O&M costs, and thus contribute to production output and effectiveness. The improvement can also permit a safe reduction in spare parts inventory and, as a result, improve capital effectiveness. Costs might be time expended and perhaps a computer application such as that to be described in Chapter 13.

Utility costs include fuel, electric power, water, and industrial gasses, as well as the cost of steam and compressed air produced centrally or within a process and distributed throughout a facility. Utility use should be metered wherever possible and not allocated. The former provides visibility and incentives for unit-specific improvements such as increased efficiency, reduced leakage, and insulation improvements.

As an example of converting value to production, a consultant brought in to survey the control air system at a large and very famous amusement park concluded that air leaks consumed the capacity of one full air compressor. In terms of net profit, air leaks required the equivalent of 10,000–15,000 added paid attendance at the park. Certainly worth identification and correction!

Regarding electric power use, recognize that electricity to run a motor-driven operating asset makes up between 50% and 85% of the lifetime ownership cost. Operating efficiency has a double impact: operating at the best efficiency point will result in lower operating costs and a longer life.

The costs of complying with safety and environmental requirements as well as any fines for noncompliance must be connected to specific units, assets, and activities where possible.

Some elements of the conversion process, such as administrative, may be strictly costs, although participants might argue that statement. Initiatives to reduce waste create value by reducing the cost of disposal.

The Value Model facilitates "what if" opportunity and sensitivity assessments, such as investment to increase effectiveness, in the only measure that counts—profit! The effect of a change in the cost of raw materials, yield, and pricing can be modeled to demonstrate a positive return within anticipated bands of variation. This capability to conduct a sensitivity analysis must be included in a value model.

A facility testing the Value Model learned that completed and planned plant, process, and equipment reliability improvements were on track to increase a performance equivalent between 6% and 8%. These improvements represent a gain of more than $100 million at the bottom line.

Value Within an Operating Environment

In addition to demonstrating the value impact of improved performance, practice, and technology within an operating enterprise, the Value Model must possess other attributes. One is the ability to predict RONA/ROCE for a given investment at any level within the enterprise and then report on value returned (results) as the investment is implemented.

Many operating managers believe that a large increase in production is more beneficial to profits than an equivalent reduction in costs. This is generally not true when comparing equivalents, bottom line results to bottom line results. Because factors such as product margins and market conditions are considered within the Value Model, these calculations will demonstrate whether increased production or reduced cost creates greatest value. This will be explored in more detail in a later section on leveraging mission/conversion effectiveness.

ROI is not nearly as effective as either a predictor or a reporter, primarily because the assumptions leading to ROI may be difficult to evaluate after the fact. In addition, conditions may change.

In a real operating enterprise, the allocation of business value to individual assets—and even components—is complicated by the existence of multiple products and the distribution of shared resources. Some may be intermediate products of another process; all require establishing internal product transfer prices. This demands an accurate division of costs between producers and users that must be factored into the ERM system.

Value Within Operations and Maintenance

O&M costs include labor costs: salaries, wages and fringe benefits, repair parts, and consumables. In addition to improvement programs directed to eliminating deficiencies, O&M itself can produce value. Effective, precision O&M practices reduce the risk of safety and environmental violations. They promote optimum reliability and direct attention to the benefits of operating at best efficiency. Operating errors and utility costs are similarly reduced by reducing waste. Waste includes out of specification product, fluid, gas, air, and heat leaks. As stated earlier, sound O&M practices also reduce requirements for replacement and spare parts, thereby improving capital effectiveness, a growing requirement in today's financial environment.

OPERATING EFFECTIVENESS

The Value Model permits tracking any given investment and determining whether the investment had the anticipated impact and if not, why not. In a sensitivity analysis, changes in the external environment such as market and price variations are considered. As stated earlier, production effectiveness is often measured in terms of OEE. An OEE of approximately 85% or greater is considered world-class performance in the discrete manufacturing industry. The number may be considerably higher in continuous process industries such as refining and chemicals.

OEE has three weaknesses as follows:

1. In terms of OEE, a process can be highly efficient—and very unprofitable—if conversion costs are excessive.
2. OEE alone does not lead to opportunity or priority. By ignoring market and business conditions, it is easy to focus OEE improvements on the wrong activity.
3. OEE can be purchased: as an extreme example, consider a facility with large design margins and installed spares for every operating asset. OEE should be quite high, as there is room for reduced performance and always an installed spare available to replace any failure. But then, consider capital effectiveness, it would be exceptionally low and never justify the initial investment.

An enterprise conducted a benchmarking study among similar industries. One finding was that larger design margins typically existed in facilities located in areas of the world where the cost of capital was low. Larger design margins resulted in greater reliability, greater tolerance to failures, and higher OEE.

For a real comparison, OEE must include the necessary expense and capital costs.

Overall Operational Effectiveness

To incorporate the crucial importance of conversion cost toward enterprise profitability, an expanded OEE-based effectiveness measure is proposed—overall operational effectiveness (OOE).

OOE adds conversion cost to OEE and modifies the availability term to consider the time window of opportunity driven by market conditions.

Timed Availability Timed availability is defined as the amount of time an operating enterprise facility, system, or component is capable of producing a required result compared to the time windows in which the mission/production is scheduled or required. Timed availability imposes three conditions to the calculation of availability:

1. For a process or facility in which the mission requires 24/7/365 operation or production is sold out requiring maximum utilization, the availability objective is 8760 h (nonleap year) to create an incentive for minimizing scheduled outages.
2. For a process or facility in which the mission is less than 24/7/365, production is not sold out, and for spared or redundant facilities, systems, or equipment, the target or objective is the actual time in which operation is required.
3. In the event that a system or component failure slows or interrupts the mission or production, the interruption does not end for the purposes of calculating timed availability until operation is fully restored and meeting all requirements. Timed availability thus reflects the full impact of a momentary malfunction, for example, electrical breaker trip that itself may be corrected in minutes yet may stop or upset the mission or production for an extended period.

Timed availability is the most realistic measure of availability for all facilities and components. It is especially valid for those that must be capable of operating at 100% during a production time increment less than total calendar time such as are experienced by peak power and food processing facilities.

Mission/Production Compliance Mission/Production Compliance is defined as the mission or means of production meeting all requirements divided by the mission/production objective. The concept of a time increment is also applied so that the term *reflects compliance* when required to meet schedule. Because actual performance can be greater than scheduled, Mission/Production compliance may be greater than 1. If off-specification production is sold at a lesser price or the mission is completed at diminished capability, a constant is applied in proportion to the reduced performance.

Quality may also be tracked as a separate quantity as in OEE. It should be noted that the Cost Of Poor Quality (COPQ) from Six Sigma provides a far more rigorous, in-depth definition that should be adopted for Operational Excellence.

Some facilities measure and track the combined timed availability and production output as production effectiveness. But production effectiveness is only part of the story. For the full picture, conversion cost must be addressed.

Mission/Conversion Effectiveness Mission/Conversion Effectiveness, the third term in OOE, is the mission/conversion cost objective divided by actual mission/conversion cost. Note that the objective is divided by the actual to reflect increasing effectiveness when actual cost is less than objective. This maintains consistency with the effectiveness terms in OEE. Mission/Conversion Effectiveness is used to measure the efficiency of a specific facility, unit, or component. All applicable costs: utilities, O&M, administrative, waste disposal, and penalties must be included.

Real Compared to Normalized Values

For a variety of reasons, not the least of which is commercial/proprietary security, some enterprises prefer normalized over real values. If so, the denominator of mission/conversion effectiveness divided by the numerator of mission/production compliance results in mission/conversion cost per unit of output, a valuable performance measure in itself. Other vital measures can be derived from OOE, provided the information structure is properly constructed.

During several discussions of OOE, participants have mentioned the difficulty of obtaining accurate cost information. Organizations must endeavor to determine costs, regardless of difficulty. They must determine exactly how much it costs to deliver a given product, comply with a mission. An ERM system mentioned earlier is a must. Lacking this knowledge, a product/mission can be easily fulfilled at less than cost—a critical and fatal mistake in today's highly competitive climate where cents per unit may be the difference between profit and loss.

Regardless of whether accurate cost information is available today, competitive survival will mandate it tomorrow. Those who cross the line between estimated and actual costs, particularly in real time discussed in Chapter 6, will have an enormous competitive advantage, as well as crucial information with which to ensure resources are always applied to highest value activities. As a final and very important requirement, if accurate historical cost and operating information isn't readily available at commencement of an Operational Excellence program, a system for recording the required data must be implemented at program commencement. Make history commence with the Operational Excellence program!

> A site approximately 2 years into a major improvement initiative suddenly realized that vital data needed to calculate value being generated from improvements wasn't being captured. Although direct participants recognized improvements to the information system were necessary, there wasn't any credible information to demonstrate value to skeptical business and financial executives. Valuable time and credibility had been lost.

LEVERAGING MISSION/CONVERSION EFFECTIVENESS

Any discussion of the necessity of linking mission/conversion effectiveness to enterprise profitability/mission effectiveness must not neglect the leverage comparison between profit increases gained through increasing conversion effectiveness (reducing conversion costs) and revenue from increased production mentioned earlier. Calculating the value from increased production must consider variable costs that may approximate 50% of the revenue gain. Variable production costs at 50% of revenue produces a 2 : 1 leverage in favor of improvements in conversion effectiveness that reduce cost. In other words, $1 million value gained through increased conversion effectiveness has the same impact on bottom line profit as $2 million of revenue from additional production. As will be expanded in Chapter 14, Finance should be able to determine a normalizing ratio between cost improvements derived from conversion effectiveness and revenue from increased production. Normalization is essential in the process of valuing potential improvements.

When availability is high and production is sold out, improved conversion effectiveness may be the only way to increase profitability.

In a calculation at a sold-out chemical plant, the profit equivalent to increasing pump average MTBF by 1 year required an availability of 103% to gain equal value.

There is also the dual contribution of increasing O&M effectiveness. In addition to the obvious advantages of reducing cost, and the leverage of increasing conversion effectiveness, there are other major contributions to value. When production is sold out, increasing output by increasing availability or run-rate contributes significantly to profit. Some companies have been able to avoid capital investment for added production by recovering as much as 40% "hidden" capacity within existing facilities.

In one case, a facility recognized one production unit was operating at an OEE of about 55%. Best practice for similar production units within their own company was 90%. Eliminating deficiencies, gaining best practice performance would gain a minimum of 35% improvement in operational effectiveness and a significant improvement in profitability.

The concept of the "Hidden Plant" originated in TPM. The "Hidden Plant" is the total production capacity lost to downtime, slowtime, startup and transition losses, poor quality, waste, scrap, and other causes. In some cases, the "Hidden Plant" may be as large as 50% nameplate capacity. More about the Hidden Plant will be found in Chapter 15.

Until presented with numbers, many executives and managers don't recognize that in reality they are operating facilities significantly below nameplate capacity.

Whatever the measurement criteria and benchmarks for mission/conversion effectiveness, all must connect directly to mission/financial objectives that are fully accepted by senior executives. Nothing else will gain support from those who control the funds. All involved with operational effectiveness must incorporate financial awareness, prioritization, and tracking of results into their everyday activities.

Without the vital dimension of business/mission/financial awareness throughout the enterprise, potentially valuable improvements in technology and practice may never be funded or applied. Furthermore, the enterprise may never perform at full potential. Instead of enlightened Operational Excellence leading to greatest value, there will be a race to the bottom—immediate and arbitrary spending reductions without any lasting improvement.

WHAT YOU SHOULD TAKE AWAY

Operational Excellence demands mission/business justification that demonstrates conclusively the real value of improved effectiveness, practice, and technology. Profit making operating enterprises must have a financial model to value prioritize improvement opportunities and track results—even when actions are separated from results by a significant time interval. The financial model must accommodate market conditions, mission, and operational requirements and demonstrate benefits in terms of financial value. The ideal financial model must contain provisions for "what if" examination of assumptions under variations in business and operating conditions. Only with the awareness provided by an accurate financial model can modern facilities be managed to optimize the only parameter that counts—profitability.

Mission-oriented operating enterprises should use the principles to construct a value model for their environment and objectives.

This Chapter is a revised and updated version of an article, **Understanding Producer Value**, by John S. Mitchell that originally appeared in the May 1999 issue of MAINTENANCE TECHNOLOGY, the magazine of plant equipment reliability, maintenance, and asset optimization, published by Applied Technology Publications, Barrington, IL. Used with permission.

6

THE ESSENTIAL EVOLUTION TO REAL-TIME BUSINESS OPERATIONAL EXCELLENCE

PETER G. MARTIN
Vice President Invensys

Through the mid-1980s, the primary business variables governing industrial operations were essentially constant for months at a time. Financial accounting systems, designed to fulfill monthly business and financial reporting requirements, categorized and totaled all the expenses accrued during the month, including raw materials, electricity, salaries, and wages. This total was subtracted from production revenue over the same period. The result: a month end snapshot of business performance expressed as Operating Profit (technically includes depreciation). Under essentially static conditions of cost and price, a separation between business/financial and operational automation systems worked reasonably well. Most operational decision-making and optimizing the operation for business drivers could be based on the monthly business report combined with a forecast of requirements for the next month.

Over the past decade, the idea of a reasonably static business environment has changed dramatically. Change in the business and competitive environments must be accommodated in near real time. The concept of profitability must be extended from a monthly report to real-time requirements at the operating level so that operations can be continually adjusted to maximize business/mission value.

Operational Excellence: Journey to Creating Sustainable Value, First Edition. John S. Mitchell.

BACKGROUND

Basing operating decisions on a monthly static financial report started to prove inadequate when key business/mission variables began to experience frequent variation. A decade ago, electricity production and distribution was highly regulated in most geographic areas of the world. With regulated supplies, industrial enterprises were able to develop contracts with their electricity suppliers for up to a year. These contracts typically set the price of a unit of electricity to a specific value for the entire year. Although some contracts included price variations based on timing and load, minimizing the cost of electricity essentially depended on minimizing consumption.

Over the past few years, electricity production and distribution has been deregulated across the globe. This deregulation has impacted different countries at different times, and not all countries have totally deregulated. However, the impact of deregulation on industrial operations has been very significant.

The pricing of any commodity is based on supply availability to demand requirements. Deregulation, opening the power grids, increased competition among electricity generation companies. As a result, the price of electricity on the open power grids began to experience much more frequent variability. As more and more suppliers linked into the electric power grid, the variability increased. The price consumers paid for electricity started to fluctuate much more frequently.

Responding to this price variability presented leading electricity consumers, such as industrial operations, with a daunting challenge to manage energy costs. A number of governments saw the issue and started to pass laws reducing the frequency of variability on the open power grids. For example, in the United States, the price charged for electricity can only change every 15 minutes. Nonetheless, this critical variable affecting the profitability of industrial operations transitioned from being a constant for months at a time to changing every 15 minutes. The system infrastructure and systems that had been developed to manage very stable business variables could no longer provide the information necessary to deal with this new, highly variable business environment.

And it is not only electricity; a similar transition is underway for other energy costs, raw materials, and production value (the current value of products produced). In the energy area within the United States, the price of gas can change every 15 minutes on the open natural gas grid. This may produce similar fluctuations in the price of energy and commodity raw materials supplied to industrial operations.

These key business variables may change many times in a single day or even during a single hour. With commodities such as metals, the degree of price fluctuation can be much faster than even energy. In the open market, the price of many base metals, copper as an example, may change multiple times a minute. The same is true for other metals. This will likely lead to major changes in the design of production plants to eliminate product and raw material storage wherever possible. New commercial channels, such as the Internet, are starting to impact the pricing variability of even consumer goods. In this real-time business environment, monthly or even daily reporting systems are insufficient for operational and business decision support driving to maximize value.

During some recent work in a chemical plant, I noted there was no raw material or product storage on site. The plant bought chemicals off a pipeline and sold them back onto a pipeline. This was unheard of 10 years ago, yet we are seeing more and more industrial companies trying to move in this direction in order to be able to capitalize on the real-time price fluctuations. This is exactly how the electric power market has behaved forever because electricity storage is very difficult and inefficient.

The fundamental differences and traditional separation between business and production control automation systems can become the primary barrier to success and gaining full benefits from the implementation of value-driven Operational Excellence. The reason for this is that the traditional thought process that resulted in the separation of the business and operational domains has changed. No longer can all business/mission functions be executed on a monthly or longer basis, because the dynamics of a typical operating enterprise have started to vary on a much shorter time frame. In addition, the responsibility for improving the profitability of operating enterprises must fall on every profit-impacting employee in the business—right down to the frontline personnel. The traditional approach to focusing on profitability at the executive and senior management levels of the enterprise; focusing on efficiency at the operating level is no longer adequate.

Much of the discussion in this chapter is directed and applicable to operating enterprises for which profitability is the essential for remaining in business. With modification, the logic and principles expressed are equally applicable to mission and service-oriented operating enterprises such as public transportation and water treating.

NECESSITY FOR REAL-TIME OPERATIONAL EXCELLENCE

Traditionally, the profitability of industrial enterprises could be effectively managed through monthly business data if the operations were running efficiently. This was due to the fact that the critical variables associated with industrial profitability were highly stable or even constant over monthly periods or longer. With a regulated, fixed price supply, reducing the overall consumption of energy over a month in an operation through efficiency-based Operational Excellence directly translates into reduced energy costs for that month. However, in the real-time industrial business environment, an industrial operation could reduce their energy consumption over the month, while their energy cost might actually increase.

Traditional approaches to Operational Excellence become inadequate in a highly variable real-time business environment. This is due to the fact that without added knowledge and discipline, it is easy to consume a preponderance of energy during high price periods. Efficiency can increase while profitability decreases. Increasing the efficiency of industrial operations no longer necessarily results in a corresponding increase in profitability. The same is true for the consumption of raw material.

As the business of industry has transitioned to real-time variability, monthly business reporting information conveying results of decisions, events, and performance that may be months if not years in the past has proven to be inadequate for making optimal business and operational decisions.

A fundamental concept of control theory is that to be able to control a variable that variable had to be measured in a time frame relative to its variability. Controlling flow might require measuring the flow every second. Controlling pressure might require even more frequent measures. When the business variables associated with the profitability of the operation did not change over monthly durations, profitability could be effectively managed with the information provided in monthly business reporting systems. But when some business variables associated with profitability started experiencing variability within daily, hourly, or even minute time frames, monthly measurement systems proved to be inadequate for effective profitability management. The monthly management systems are still required for financial reporting but have become ineffective for decision support.

Real-time control of the profitability of industrial operations has become a very important aspect of modern Operational Excellence programs. Effective Operational Excellence in the emerging real-time industrial business environment must be directed beyond efficiency to profitability. Business and operational systems must be totally integrated through Operational Excellence to have the desired positive impact on business value. Business executives are interested in profitability. They are interested in efficiency only when it leads to increased profitability. Profitability improvement must become the job of every person in the operation.

INTEGRATED BUSINESS AND OPERATIONS SYSTEM

Addressing Operational Excellence involves much more than merely connecting business systems to operations systems. Business systems have been optimized to report business results over a time period in the past, not to provide anticipatory operational decision support. Historical reporting is an essential function of a business and should be considered such. But improving the performance of the business requires effective business decision-making in real time or near real time right down to frontline operating personnel.

For decision support information to have impact, it must be provided in the right format, to the right place, at the right time. This is not the case with today's business reporting systems. It must be the case for the Operational Excellence system. In essence, each person who might impact the performance of the operation needs to become a business performance manager in the business domain for which they are responsible. This is the new challenge that business executives are working to make a reality. The real-time business decision-making approach must be incorporated as part of the next generation of Operational Excellence.

Real-Time Business Focus

Incorporating a real-time business focus into Operational Excellence presents a daunting challenge. A quick review of any industrial operating enterprise's annual

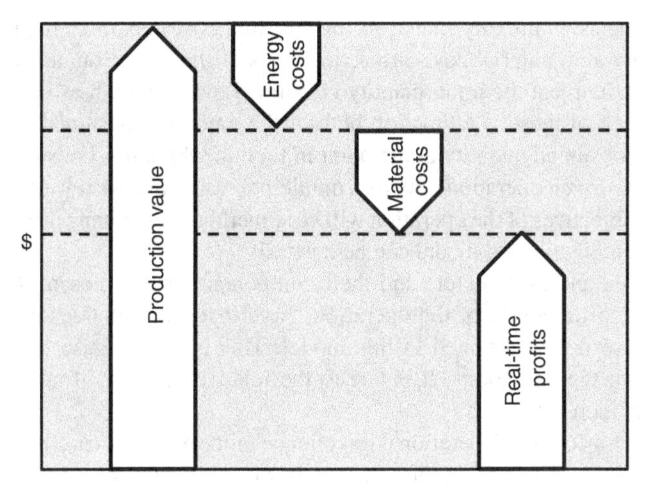

Figure 6.1 Constrained real-time business-focused Operational Excellence profitability model.

report reveals the vast quantity and variety of information reported on business performance. Having to deal with all of this information in real time can be overwhelming. The good news is that the business variables that fluctuate in real time are a very small subset of the data in a financial report.

Looking at the static financial model presented in Chapter 5, the components of profitability that may fluctuate on a sub-daily basis are production value (sold output times selling price), energy costs, and material costs. Thus, with some modifications, the model shown in Chapter 5 can be shifted into a real-time representation.

A simplified, constrained, real-time profitably waterfall model is shown in Figure 6.1. Production value, value of products produced, is a primary component of profitability. It is displayed as the first bar in the diagram. The material and energy costs are shown as two reverse direction vectors that reduce the profitability from the production value vector. The final vector represents the real-time component of the profitability of the operation. Each vector is made up of two components. As stated in Chapter 5, revenue from sales consists of sold production output multiplied by selling price [for the revenue calculation, the production output quantity may require adjustment when less than 100% of production output is sold (transfers to inventory)]. The two cost vectors consist of two multiplicative quantities, value of energy or material consumed times the amount consumed.

Maximizing profitability, the final vector is the primary real-time business objective of all profit-driven industrial operations. All four vectors vary in real time, which makes controlling the real-time profitability of the operation a very daunting challenge. For the purpose of this discussion, it is assumed that production is operated and output sold to a forecast demand and selling price is determined outside the control of Operational Excellence. Thus, production output in compliance to business requirements is one principle objective of Operational Excellence; the other is maximizing cost effectiveness. Optimally managed revenue and cost maximize the profitability vector. Energy and material costs are actual costs of the operation, as they are consumed rather than just the total quantity consumed over a month as in the financial statement. Each of these is a function both of the amounts consumed and the price paid for the consumed quantity at the point in time in which it is consumed.

For budget-driven operations, such as municipal water and wastewater operations, the business objectives of the operation will be something other than pure profitability, but a similar, budget-based model can be derived.

The revenue and cost vectors and their component parts represent the real-time aspects of the profitability of the operation. Note that other obvious costs, such as labor costs, are not represented in this model. This is not because labor costs are unimportant to the operations. It is due to the relative stability of labor costs over monthly time increments.

This business-focused Operational Excellence model requires careful reconsideration of the functional scope that must be incorporated into the Operational Excellence domain to have the best chance of meeting the objective of maximizing profitability (or whatever the primary business/mission objective may be). Every key function, department, activity, and task that impact profitability must be incorporated within the scope of business-driven Operational Excellence. This must include enterprise and plant management, operations, maintenance, work planning and scheduling, engineering, environment, safety, plant finance (accounting), and IT. To gain maximum value, it is imperative that these functions collaborate effectively. Most productive innovations occur when multifunction teams work together to solve problems. Decision-based collaboration is the norm for Operational Excellence. Repeating a quotation from Chapter 1 for emphasis:

> "Employee led leadership teams do unbelievably good strategic and tactical planning—if you give them the opportunity. Most important, they gain total buy-in for the plan, its implementation and results."
>
> Fortune 250 company CEO

The challenges associated with decision-based collaboration cannot be overstated. In the first place, functional domains were initially established due to an affinity between the complexity, knowledge, and skills required in each area. Each function requires very specific education and experience, and each has its own unique vernacular. The knowledge required to be successful in one area is completely different from that required for the other functions. Very often, even the words used in one are semantically disconnected from those used in other functional domains. To exacerbate the situation, over the years, the people in one functional area may have grown to distrust or dislike those in others. These organizational issues may actually present the largest single obstacle to the implementation of an effective business-driven Operational Excellence strategy.

> Cooperation between functions at all levels is absolutely essential to gain maximum performance and effectiveness. No one has all the answers—together we do.

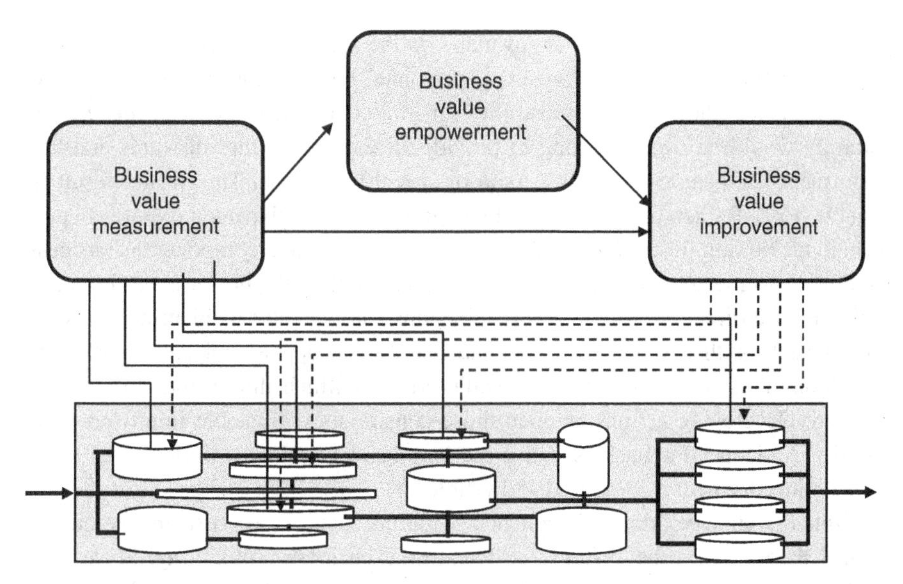

Figure 6.2 Business value-focused Operational Excellence—real-time profit control system.

DEVELOPING REAL-TIME BUSINESS-DRIVEN OPERATIONAL EXCELLENCE

The implications of business variables transitioning to real-time variability are significant. Industrial operations that had invested in efficiency-improving activities and programs have started to realize that although their efficiency may have improved, their profitability may not if reduced use gained through improved efficiency is cancelled by increases in the price of energy. Monthly business information generated by Enterprise Resource Planning (ERP) systems isn't sufficient for effective operational decision support. The business measurement systems do not synchronize with real-time business dynamics.

Approaching business-driven Operational Excellence represents a classic real-time control problem. In this case, the focus of the control goes beyond the operating processes to incorporate business processes. As with any other real-time control problem, there are critical control components that must be developed to bring the profitability of the operation under control and to drive continuous profit improvement. Assuming that there is an effective system controlling the efficiency of the operation, a higher level control system is additionally required for the real-time components of profitability. As with plant level operational control systems, there are three basic components that comprise a profitability control system, Figure 6.2. The first component is based on measuring the business variables to be controlled in real time. Measuring the key business variables that need to be controlled is the essential enabler to make control possible. This is a basic concept in control theory, but one that is often overlooked when developing an Operational Excellence strategy.

Once the business measures are available in the appropriate time frame, the next step is empowering all profit-impacting personnel by providing them with a view of the measures they have responsibility for in a context that they can relate to. It certainly would be overwhelming to provide all the accounting measures and key performance indicators to every person on a real-time basis. This would result in data overload. Rather, providing exactly the information each person requires to perform in his/her job function is much more effective. The final component is business value improvement—performing specific actions to drive the business performance in the desired direction. The appropriate actions can typically be identified through a bottleneck or theory of constraints analysis for the operational area under consideration. This will help identify specific areas in which the profitability is constrained. Developing actions to open those constraints will enable improved profitability. And since the business value measurements are already in place, the value of each improvement is measured and visible.

Value-driven Operational Excellence solutions require expertise in operations, control theory, real-time systems, and real-time business profitability. Since they understand and work in the first three areas, operations, maintenance, and engineering, professionals must be in the lead with support from business/accounting and IT. What all tend to lack is a solid understanding of real-time profitability. However, since the transition of operational business variables to real-time fluctuation is a reasonably recent trend, most industrial enterprises may have to select innovative professionals and develop this expertise in-house.

Developing business-driven, real-time, profit-based Operational Excellence requires the application of control theory to the real-time business variables of production value, energy cost, material cost, safety risk, and environmental risk. Certainly, each of these real-time business variables can be controlled independently much in the same manner as flow, level, temperature, and pressure are controlled within the production process. But since the real-time business variables are not independent (as is clear from Figure 6.2), it is best to develop a control strategy that coordinates these business variables in a holistic manner in order to achieve the optimal balance between them that safely gains maximum profitability over time at acceptable risk levels.

To date, industry has very little overall experience controlling real-time business variables. Therefore, the best way to develop a coordinated and balanced approach is to use the talent in the enterprise through a real-time empowerment strategy. It is interesting to note that industrial engineers have yet to develop an effective mathematical approach to resolving real-time multiple objective optimization, but an experienced operator with a the equivalent of a high school education and effective empowerment tends to be able to learn how to balance these variables for optimal profitability over only a few weeks. Effective business-driven Operational Excellence requires industrial organizations to take full advantage of all personnel and to transition every profit-impacting person in their organizations into real-time performance managers.

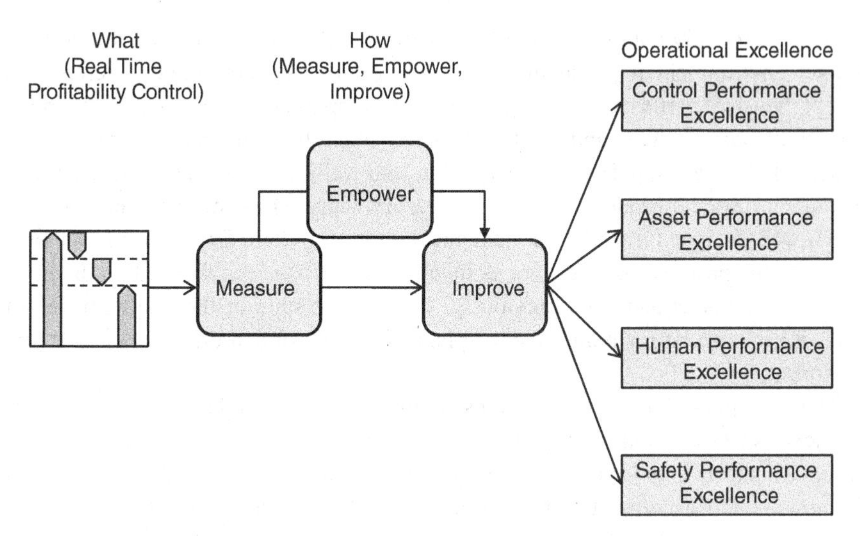

Figure 6.3 Business-focused Operational Excellence.

Segmenting a Complex Concept

As mentioned in Chapter 1, experience has shown that trying to tackle an overall Operational Excellence strategy as a single project can be highly complex and extremely daunting. Thus, partitioning the strategy into a few focused areas reduces complexity, engages the talents of the enterprise most effectively, and thereby greatly improves the probability of success. Figure 6.3 overviews the critical aspects of a business-driven Operational Excellence process, breaking the execution into four key areas: (production) Control Performance Excellence, Asset Performance Excellence, Human Performance Excellence and Safety Performance Excellence. (Note: this is a slightly different and more specialized view from that outlined in Chapter 3 where Human and Safety Performance Excellence are viewed as common elements that apply to all functional divisions including several not mentioned in this chapter, Chapter 8.) Although the areas are highly interrelated, experience has shown that they can be approached somewhat independently if each is based on the overall business objectives of the operation. There must be a single control and administrative structure, with efforts effectively led and coordinated by a single Steering Team within an overall Operational Excellence strategy.

Control Performance Excellence The first of the functional areas within the business-driven Operational Excellence model is control performance excellence. Control performance excellence encompasses the plant floor control and is initially directed to improving and maintaining the efficiency of the production process. It includes the business control focused on maximizing profitability determined by production efficiency and optimal cost illustrated in Figure 6.1. Although improved profitability is determined by both improved efficiency and the time of usage as mentioned previously, the process begins with improving efficiency. In fact, it is impossible to truly control the profitability of the operation in the current real-time market environment if production efficiency is not well controlled. Therefore, a comprehensive performance excellence control approach will start at the plant floor with an effective and well-maintained plant floor instrument and control system. This must include accurate, reliable, and well-calibrated instrumentation for manufacturing and production process measurements. These must be incorporated into a well-designed, tuned, and maintained control system. Experience has shown that over the past few decades, not as much attention has been focused on the plant control systems as had been previously. The current state of these systems is not what it needs to be for both effective plant floor control and effective profitability control.

Once the plant floor control systems are up to the essential level of effectiveness, the next aspect of control performance excellence is implementing real-time profitability control. This is accomplished in a very similar manner to the plant control systems. The measurement of the real-time profitability variables must be effectively and accurately developed right down to the plant, unit, or work cell level. This can be accomplished by modeling the profitability measures in the real-time control environment using plant instrumentation as the primary real-time input to the models. Transfer pricing for stages along the production process are essential. The static model presented in the previous chapter will be of help.

The plant instrumentation transmits signals that convey what the plant is doing second-by-second. This information can be transformed into real-time profitability measures by combining it with other, less real-time business information either from the ERP system or from other external data sources. These measures can be used to empower plant personnel by providing prioritization for each person driving value in the operation. The prioritization should be performed according to the currently executing manufacturing strategy of the operation. Although profit-based empowerment will certainly help plant personnel to continuously improve profit, specific initiatives can be identified and implemented to further improve profitability through a structured theory of constraints analysis on the profitability within each plant unit and across the plant as a whole. The result will be a profitability control system.

The output of the profit control system will provide guidance to the plant control system. In a control theory sense, this results in a cascade control system with profitability control as the primary loop and plant efficiency control as the secondary loop, Figure 6.4. In today's dynamic business environment, external business and operational conditions are bound to vary over time. Changes in external market conditions need to be incorporated in the profitability control system, which is the primary loop of the profitability to efficiency cascade control scheme. The static model illustrated in Chapter 5 has inputs for variations in market conditions and pricing. External changes impacting operational variables must be likewise incorporated in the efficiency control system, which is the secondary loop of the scheme, as has been done for many years. With this holistic control structure, both real-time efficiency variables and real-time profitability variables can be effectively controlled.

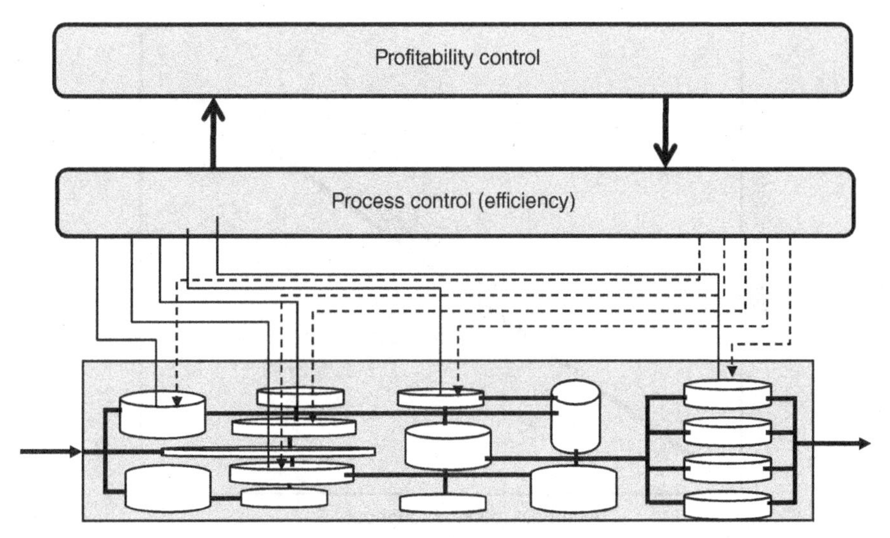

Figure 6.4 Controlling profitability and efficiency in real time—cascading profit to process control.

Asset Performance Excellence A second functional area of Operational Excellence is asset performance excellence. The idea of asset performance excellence began in the late 1990s with Physical Asset Management. As originally conceived, Physical Asset Management emphasized reliability and minimizing risk. It included the working culture, human, and organizational elements considered essential to gain maximum sustainable performance and value from physical production systems and assets. Asset Management promoted processes and technologies that would significantly improve the efficiency of maintenance and thus was received positively in industry. Since its introduction, asset management has evolved to become essentially synonymous with maintenance. Largely lost were the stated necessity and essential contribution of reliability, risk mitigation, a value- and improvement-oriented working culture including multifunction teams to identify and resolve problems, refined processes, and procedures.

With time, it has become clear that maintenance of physical production assets is only one aspect of asset effectiveness. Inherent reliability and the way in which an asset is operated have a significant impact on both the performance and the maintained state of the asset.

In many industrial operations, the maintenance organization and operations team are managed as almost completely independent functions. Exacerbating this separation, the performance of the maintenance function is typically measured on asset availability and cost compliance to budget. Performance of the operations team is typically measured on throughput; production rate. In this relationship, production rate and asset availability tend to be inverse functions (at least partially). This means that increasing production rate typically results in some reduction of availability and/or increase in maintenance costs. Increasing asset availability,

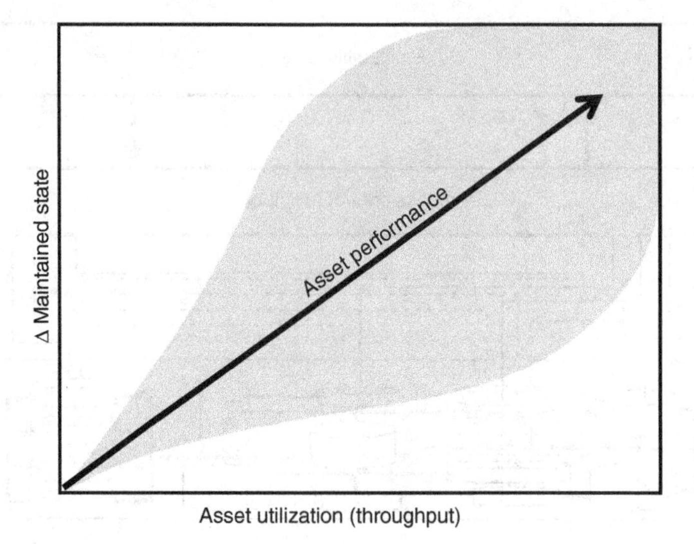

Figure 6.5 Asset performance excellence—combining maintenance and operations to gain maximum financial value.

reducing maintenance costs might be accomplished by reducing production rate. This inverse relationship tends to drive a wedge between the maintenance and operations teams and creates a noncooperative environment. It also tends to suboptimize the performance of the assets.

As mentioned in the previous chapter, and often more important than inverse performance measures, revenue from production throughput is the top line of a Profit and Loss statement of financial performance; cost essentially impacts the bottom line directly. Thus, cost has a multiplier effect that can be as large as a factor of 10 compared to production revenue when translated into the business domain in terms of operating profit/loss. Increased throughput achieved at a cost of additional maintenance must be modeled to assure the most profitable decision.

A primary aspect of the solution to this conflicting situation is the development of a combined performance measurement system that drives collaboration rather than conflict. Experience has shown that such a mediating performance measure could be asset performance, defined as the amount of business value driven through the asset both in real time and over time. The static model illustrated in Chapter 5 could serve as a starting point. From another perspective, driving the asset harder to produce more output negatively impacts the maintained state of the asset which will, in turn, negatively impact the asset availability and cost over time as shown by the shaded region in Figure 6.5.

The goal of the operation should be to operate and maintain the asset in a manner that maximizes the value from the asset over time, represented by the "Asset Performance" vector in Figure 6.5. Measuring the maintained state of any industrial asset is challenging and often requires a sophisticated dynamic asset modeling approach based on condition measurements. However, measuring the performance of each asset has proven to be more straightforward and is basically accomplished from control system variables.

Combining real-time, or almost real-time, mechanical reliability, risk, and condition variables with operating performance provides an excellent collaborative measure of performance for both operations and maintenance. Time variations and trends, can be used for a good estimate of future condition. Considerable improvement can be realized by focusing on this combined measure rather than worrying about modeling maintained state. By creating an overall operations and maintenance performance measure and trending it over time for each asset, common goals can be developed between the two teams and a collaborative spirit developed. Most productive innovation occurs when two or more different groups join forces to solve problems. This collaboration is promoted and facilitated by an Operational Excellence program and typically results in very innovative and constructive combined solutions that create greatest value. The result is the continuous improvement of the operational profitability generated through the asset base whether fixed or mobile. This is the essence and the primary benefit of Operational Excellence.

WHAT YOU SHOULD TAKE AWAY

Strategies similar to Operational Excellence have been executed in industrial companies for the better part of the last two decades. Recently, there has been a significant transition in focus. The initial emphasis was almost entirely on improving plant operating efficiency. This initial focus was appropriate in the period in which industrial operation business variables were highly stable over long time frames. Over the past few years, some of the key business variables have begun to transition to real-time variability. Under these conditions, the traditional approach to performance efficiency as a measure of profitability has proven to be inadequate. Industrial enterprises still completely focused on efficiency improvements are finding that the profitability of their operations is starting to get out of control.

Pursuing an Operational Excellence strategy by focusing on control performance excellence, asset performance excellence, human performance excellence, and safety performance excellence (safety and human performance excellence are covered in Chapter 8) can allow industrial organizations to mobilize their talent most effectively within a holistic structure. These four components of Operational Excellence are not at all independent from each other. As industrial organizations start focusing on improvement opportunities and results, they will find that the four areas are mutually supportive on many of the same issues. The multifunction improvement action team that is the center of Operational Excellence reconciles functional issues as the more holistic Operational Excellence is executed.

Reconciliation is accomplished by initiating each subcomponent program with a top-down analysis of the enterprise and production objectives and strategies, action plans, and resulting performance measures. In this way, each of the subcomponents solutions will be inherently aligned to the strategy of the operation. After each subcomponent solution is implemented, a final solution phase must be undertaken that intentionally converges the solutions into a common holistic solution aligned to the strategy of the operation. This tends to be fairly straightforward to accomplish once the subcomponents are each similarly aligned.

This new approach to Operational Excellence expands the traditional functional boundaries to incorporate profitability control as well as efficiency control. It is necessary for continuous profitability improvement, the ultimate objective of business management.

7

LEADERSHIP, VISION, STRATEGY

Similar to so many, if not most, initiatives, the success of Operational Excellence depends on people. Visionary, motivated people who have initiative and commitment, accept ownership for the program and responsibility for success. The necessity for commitment begins in the executive suite and extends throughout the enterprise all the way to the working level. Recognizing that not all people in an enterprise fit this profile, success requires selecting the very best of those who do to lead, develop, and participate in the Operational Excellence program. Enthusiasm and success are force multipliers. They will encourage and bring the skeptical and reluctant into the fold as the program progresses and gains success.

A recommended organization of the Operational Excellence program is illustrated in Figure 7.1.

EXECUTIVE CHAMPION

To gain full success, the Operational Excellence program must begin with a strong, visibly committed, executive sponsor/champion. The Executive Champion must be at a high level within the enterprise with sufficient power to approve improvements to the administrative organization, practices, and technology. Promoting success and eliminating barriers are essential to gain maximum sustainable value.

The Executive Champion will be referred to in the first person. Realistically, the Executive Champion will likely appoint a staff person to act in his/her stead. If this is the course taken, the Executive Champion must choose an experienced, highly competent, respected individual who has his/her full confidence and make it clear that the deputy is fully empowered to act and make decisions. The deputy must be willing

Operational Excellence: Journey to Creating Sustainable Value, First Edition. John S. Mitchell.

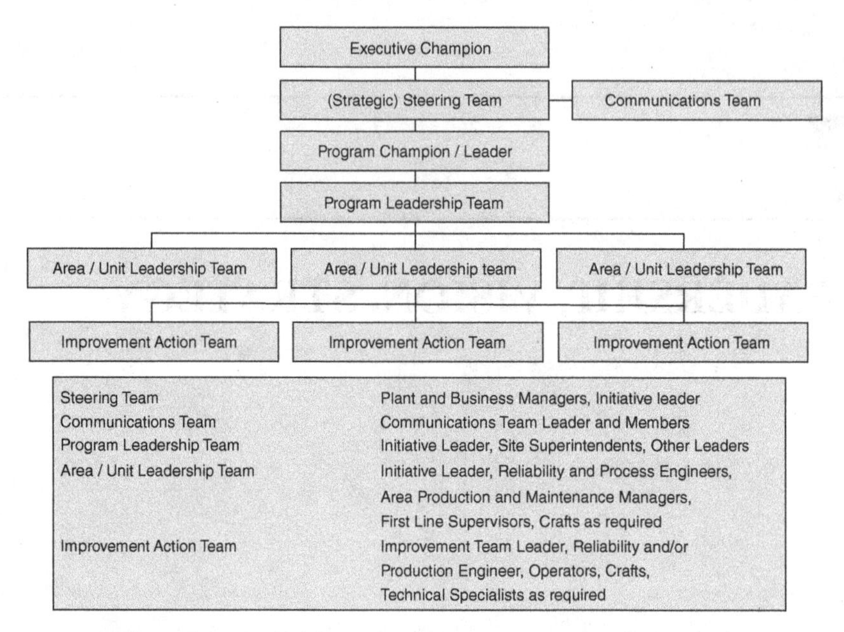

Figure 7.1 Typical Operational Excellence program organization

to make most decisions on the spot, safe in the knowledge they will be supported fully by the Executive Champion. Stating that "I have to pass it by my boss" diminishes the deputy, slows the process, is contrary to the empowerment principle of Operational Excellence, and is a negative example to all in the program.

Even when the Executive Champion delegates authority for the program, he/she must maintain contact and awareness of what's transpiring, success, challenges, and barriers. The Executive Champion must take the time to issue communiqués that convey knowledge, interest, and drive even if the communiqués are written by the designated deputy. During site visits, the Executive Champion must visibly convey interest and commitment by conversing with program leaders and participants. Although many of the duties of the Executive Champion can and will be delegated, he/she cannot delegate vision, continuing genuine interest, and drive that are essential for program success.

Visibly Engaged

The Executive Champion must be a visionary, enthusiastic, results-oriented individual who sponsors, energizes, advocates, and actively drives the initiative. The Executive Champion must be fully supportive, enthusiastically engaged, and visibly committed to success. He/she must be observed asking questions and commenting on the achievements of participants, thereby demonstrating up-to-date knowledge of the program, participants, successes, and challenges. Since motivation is all important, conveying interest and leadership "by walking around" is imperative.

> A normally reserved site manager who stated his commitment to an improvement initiative but had displayed little or no personal involvement was encouraged to demonstrate commitment by occasional participation in working meetings. During the first meeting he attended, and after some reservations, participants began asking questions. Some were very direct and tough. From the answers, it was quite apparent the site manager knew exactly what was going on in the program, was highly supportive, and pleased with results. The interchange provided a great boost in the participant's morale and commitment for the program!

The necessity for engagement and drive from the Executive Champion cannot be understated. Commitment must be real, continuing, and visible and is one of the most, if not the most, essentials for success. If the organization perceives the top is absolutely committed with clear objectives, chances are they will follow.

Communicate Compelling Vision

Vision: an idealized, collective image of an objective, future state.

The Executive Champion originates and communicates a vivid description of the necessity and benefits of the journey to excellence. The executive vision must be a compelling description of where we have to go and why. The vision must describe a future state that will gain enthusiastic support for the journey to operational excellence and promote motivation and ownership at all levels of the organization. Ideally, the vision must define the business necessity for improvement, why improvement is necessary, what must be accomplished, and emphasize positive benefits for all.

The vision for the future state must be quite clear with very specific and defined measurable objectives, published, understood, and accepted. Following are three vision statements with these characteristics:

> "Safely gain a position within 10% cost per unit output of the lowest cost producer in our industry within 5 years."
>
> "Achieve market price, production capacity factor greater than 70%, equivalent availability of 90%, improve heat rate by 5%, and reduce O&M costs to 3.5 mills per kWh."
>
> "Ten by ten: reduce production costs by $10 per unit by 2010."

The next vision statement, paraphrased from the website of a major U.S. Corporation, was published during a downturn in which price pressures were increasing and demand was decreasing:

> "We must make changes to more tightly align our manufacturing with market and customer requirements and reform our cost structure for greater consistency with business and market realities."

It was not received positively. Employees reading the vision statement correctly concluded that layoffs were in the offing. As a result, the most competent employees who would have been retained under the most severe circumstances, and who also were the most desirable to other employers, immediately began circulating resumes. The published vision statement produced an opposite reaction from the intent.

Define Overall Enterprise Business Strategy

The Executive Champion must communicate the overall business/mission strategy, including ambitious, optimistic objectives, with all the necessary information to direct the Operational Excellence program. Is the business/mission strategy growth, increasing efficiency/effectiveness, or a combination of the two? Refer back to Figure 2.1. The three vision statements noted previously are all directed to increasing efficiency. A growth strategy could be to gain market share.

> A company with completely sold out production had as a mission to increase production output by 20% without any increase in fixed costs.

This mission is a combination of growth and increasing efficiency.

Translate Business/Mission Objectives into Program Requirements

The Executive Champion has the initial lead to interpret enterprise business/mission objectives and state broadly how they should translate to program objectives. Are the principal objectives of Operational Excellence improving mission compliance, efficiency, production output, reducing costs, or some combination? Note the preceding vision statements; all are very specific and will lead to the improvements and activities necessary for compliance.

In all cases, the strategy must clearly state that success in a subordinate improvement program such as Operational Excellence is directly dependent on fulfilling enterprise mission/business objectives.

Select, Appoint, and Empower Operational Leaders

Selecting and empowering strong, committed operational leaders is a task of the Executive Champion that has far-reaching consequences. This begins with the Steering Team, to be described in a later section. Operational leaders must be highly respected within the organization, represent all the major functions included within Operational Excellence, and have the prestige and power to facilitate activities and remove functional or other barriers that may exist or arise.

Appointees to leadership positions must be committed to success, willing to work across organizational lines, and occasionally sacrifice what might be considered a functional imperative for the overall good of Operational Excellence. In short, leadership appointees must be genuinely and visibly committed to the success of Operational Excellence and the Operational Excellence program.

Continuing Tasks

The Executive Champion has ongoing tasks, primarily oversight and encouragement, as the Operational Excellence program develops and matures.

As stated, assisting in the development of program-specific objectives and the strategy to accomplish the objectives is one of the first. It is imperative that the program's strategic direction and specific objectives make the greatest contribution to enterprise mission and business objectives as soon as possible. This requires quickly identifying specific tasks and activities that will produce maximum value and benefit to mission/business objectives in the least time. In many cases, even the best operational leaders are not accustomed to thinking in terms of mission/business value. The transition and change in mindset must be facilitated by the Executive Champion, so all are working toward the same objectives.

As the Operational Excellence program is organized and gathers momentum, the Executive Champion has a continuing role. From demonstrating total commitment to the program and its success, overseeing and approving program-specific objectives, strategy, and improvement initiatives, to monitoring progress and results, encouraging, empowering, facilitating and congratulating success, the Executive Champion must take the broad view. The Executive Champion must be results oriented, monitoring program progress frequently to assure results are being produced that meet expectations for mission/business value improvement and commitments of the people involved and communicating congratulations for success when attained. At the same time, the Executive Champion must be patient, allowing the program to continue so long as progress is being demonstrated, recognizing that results take time and sustainability longer yet—*Rome wasn't built in a day!*

Operational Excellence is a continuing journey, not a program of the month.

As the Operational Excellence program matures, re-examining the program strategy and plans to assure continuing alignment with the enterprise business/mission objectives and strategy, as well as maximum contribution, is a necessity.

It bears repeating that authorizing Operational Excellence is just the first step. The executive champion must remain engaged, promoting and driving the program with visible encouragement. Executives must assure adequate resources in the form of funds, people, and training/skill development sufficient for the program to deliver objective results as defined in the business/mission strategy. Commitment flows top-down; the executive champion must be fully and visibly committed to expect equivalent commitment and ownership at the working level.

STEERING TEAM

As stated in an earlier chapter, it is imperative that all site/facility improvement programs are brought together within a single, master Operational Excellence program. A single Steering Team provides top-level strategic leadership and direction to assure common objectives, complementary plans, and coordinated efforts across all improvement programs under the roof of Operational Excellence. The Steering Team champions, drives, actively leads, facilitates, and coordinates the program. Equally important, the Steering Team monitors progress and assures results.

The Steering Team is composed of the best, most respected and highly motivated operating executives and functional managers. A Steering Team typically has approximately 7–10 members. There can be more, and there may be less. It is imperative to have the best, most highly motivated executives and senior managers from Operations/Production, Maintenance, Engineering, Safety, Finance, Information Technology (IT), and Human Resources (HR) as members of the Steering Team. Others may be added to assure all functions involved in the Operational Excellence initiative are represented by high level participation. Team members must have keen ownership for the program and total commitment to success. They must possess the personal power and willingness to resolve most intra- and extra-function conflicts. The site/facility manager is an ex-officio member.

Steering Teams generally function best if the senior operations/production executive heads the team. Other arrangements can be successful depending on the specific participants, position and respect within the organization, and ability and time to lead a major improvement effort and gain cooperation throughout the organization.

General Guidelines

Members of the Steering Team must recognize and keep uppermost in mind that the Steering Team is a strategic, not an operational, tactical entity. As such, the primary role of the Steering Team is to provide strategic direction, champion, drive, actively lead/facilitate the program, monitor, and assure results. The Steering Team should concentrate on strategic issues that will contribute to program success. Most importantly, this includes identifying and translating business objectives into program objectives and specific results required. Assuring the program objectives are directed to enterprise mission/business objectives, prioritized to achieve greatest value and provide measurable contribution is a continuing task of the Steering Team. Facilitation to assure teamwork and cooperation between functions is likewise a key role of the Steering Team.

The Steering Team should limit their involvement in tactics to overseeing and facilitating the working-level improvement opportunity identification, planning, and implementation process. The Steering Team reviews improvement action plans, validates potential business value, risk, and actions to gain objectives. When satisfied, the Steering Team approves improvement plans including intermediate and final results metrics, time lines, and resources required.

It is not the task of the Steering Team to develop detailed improvement action plans. The Steering Team should agree that, with oversight, improvement initiatives will be identified and developed by those who will have direct responsibility for implementation. This creates ownership and responsibility and is imperative for success.

There is one tricky bit in the approval process: most developers of an improvement initiative will be highly enthusiastic for their plan and want to regale the Steering Team with all the wonderful technical details during the approval process. To demonstrate that the individual/team has full responsibility for success, the Steering Team should develop a procedure for approving improvement action plans that includes a time limit for the presentation and specifies the information to be conveyed, for example, improvement objectives, future state, basic steps, value contribution, metrics including time line, changes to any established processes or procedures, resources required, and risk. Risk includes the probability the plan will produce desired results, consequences if it doesn't, and mitigating actions to reduce risk.

The Steering Team should have decision authority to provide most resources required for an improvement plan. Post implementation, the Steering Team serves as a facilitating resource, energizing implementation, assuring the organization is fully aligned, and working toward success. This includes improving cooperation and removing organizational barriers to success.

The Steering Team monitors and periodically reviews the function, progress to objectives, and results of the various improvement initiatives, including improvements necessary to meet objectives. Where results are underperforming expectations, the Steering Team should make recommendations for improvement. Here again, final responsibility rests with the team directly responsible.

Appoint Operational Excellence Program Leader/Champion

Appointing the leader/champion of the Operational Excellence initiative is the first task of the newly formed Steering Team. Attributes of the position will be described in a later section. The Operational Excellence program Leader/Champion becomes a member of the Steering Team.

Define Program Charter

The next order of business for the Steering Team is to formulate an initial draft charter providing overall guidance for the Operational Excellence program. Full requirements of the program charter will be covered in detail in Chapter 9: the Define stage of the Operational Excellence program.

For the purposes of this overview, the program charter begins by translating the overall business and mission objectives into program-specific objectives and overall strategy for attaining the objectives. Much of this task should have been completed by the Executive Champion. If so, the steering team reviews program objectives and strategy, making refinements as required to meet enterprise business/mission objectives. The charter defines the Operational Excellence mission, results required, and the strategy to gain the results. The mission and strategy provide specific guidance as to how the Operational Excellence program will fulfill enterprise mission/business objectives and strategy. Guidance includes prioritization in areas such as improve mission/production effectiveness (OEE/OOE; overall equipment effectiveness/overall operating effectiveness) and reduce waste and cost. The preliminary draft might contain suggestions for specific improvements. It should not get into details. Details will be developed by Program Leadership and Improvement Action Teams to be formed during the program Define and Plan stages.

A preliminary mission statement is the next task of the newly formed Steering Team. The mission statement must be definitive with measurable objectives. Development, review, and finalization of the mission statement will be covered in more detail later in the description of the program Define stage.

The charter should include a proposed administrative organization for the initiative and preliminary roles and responsibilities (RASCI), as shown in Figure 7.2.

Operational Excellence program	Executive Champion	Steering Team	Program Leader	Leadership Team	Improvement Team Leaders	Communications Team	Site / Plant Manager	Site / Plant Management	Support Personnel
Establish program necessity, vision, business objectives	A	I	I	I	I	I	I	I	I
Translate business objectives into program objectives	S	A	R	I	I	I	I	I	I
Convey program necessity and benefits to site personnel		S	S	C	C	A	S	C	C
Develop risk analysis and value prioritization procedures		A	R	C	I		C	C	
Develop program strategy and refine objectives		S	A	R	C	I	C	C	C
Identify and prioritize improvement opportunities		C	R	A	R	I	C	C	C
Develop improvement action plans		I	R	S	A	I	C	C	C
Approve improvement action plans		A	R	R	R	I	S	S	S
Implement improvement action plans		I	S	S	A	I	S	S	S
Follow up to assure results conform to objectives		I	R	S	A	I	I	I	I
Publicize successes	I	S	S	S	S	A	I	I	I
Identify additional opportunities for improvement		C	R	R	A	I	C	C	C

Figure 7.2 Operational Excellence responsible, accountable, support, consult, inform; RASCI diagram

A preliminary process for reviewing and approving improvement initiatives should be outlined.

Finally, the Steering Team needs to review the enterprise and site methodologies for assessing value and risk. If there isn't a definitive methodology for either that satisfies all internal, external, and standards requirements, for example, financial model, as mentioned in Chapter 5, ISO, etc., they must be developed and approved.

Facilitate Teamwork

Facilitating organizational teamwork within a multifunction program in which many won't be comfortable is an initial and continuing task of both the Executive Champion and Steering Team. Achieving cross-functional teamwork and often teamwork in general may require effort and training. Some have found that team training is a beneficial beginning for people who may be unaccustomed to working collaboratively with others. At the beginning of the Operational Excellence program, the Executive Champion, Steering Team, and Program Leader must all be sensitive to the requirement for team training to establish positive and constructive initial conditions.

> People working together constructively in teams to accomplish a common, agreed upon goal can accomplish significantly more than the same individuals working alone. A good team is greater than the sum of its parts.

Even with the world's greatest teamwork, there may be organizational and institutional barriers to success. These must be identified as early in the program as possible and rectified. During the initial stages, the Executive Champion must be very sensitive to organizational and institutional barriers, especially within the Steering Team where members may have a past history of competition or treating each other as superior/subordinate. If this occurs, the Executive Champion must act quickly and decisively to achieve resolution. As the program matures, there will be fewer and fewer barriers requiring attention from the Executive Champion.

Empowered Decisions

Empowerment throughout the Operational Excellence organization is a key to success. The Steering Team must be empowered to make most program decisions. Similarly, within clear guidelines mutually established between the Executive Champion and Steering Team, working-level leadership and improvement teams should be empowered to make decisions. Areas to consider in empowering teams and individuals to make decisions should include an evaluation of worst-case risk, and what could happen if following the decision, tasks, activities, and results don't turn out as planned.

> Participants in Operational Excellence workshops often complain that many simple decisions that have to be made and have little or no consequences require higher authority. The requirement diminishes their status and implies they are considered less than

> responsible. One individual cited a requirement in his organization that required repair parts over a certain monetary value to be signed for by a division manager. He stated that there wasn't any question the repair part was required. The work was authorized by a Work Order, and custody was established by the working-level person who needed the repair part. Adding to frustration and a feeling of distrust, the division manager who ultimately signed the request often had no idea what the repair part was for or what it did.

As a part of the empowerment process, the Executive Champion and Steering Team need to agree that with guidelines and oversight, improvement initiatives will, for the most part, be identified and developed by those who will have direct responsibility for implementation. This is crucial to establish ownership, responsibility, and accountability for results. Much more difficult to make excuses for something that was mine from the beginning compared to yours!

> The opportunity discovery process and development of improvement initiatives are essentials to gain the energy, ownership, commitment, and responsibility necessary for achieving maximum results and sustainable success.

LEADERSHIP SUCCESSION

Recognizing that Operational Excellence is a continuing journey, there must be a leadership succession plan agreed on and in effect. This will assure program continuity and minimize the possibility of detrimental changes if people in leadership are replaced for any reason. Leadership includes the Executive Champion, members of the Steering Team, and the program leader.

> All too often promising programs are changed for the worse or abandoned altogether when key leadership personnel change. The incoming leadership may not have any background or vested interest in the program, may not consider it "theirs," or feel they will receive credit for success. The combination of lack of ownership, together with a conclusion that the incoming leadership won't be credited for success, often spells the end of a promising improvement program. Most people who have been with an enterprise for an extended period can cite several instances where a program that was just beginning to show results was terminated under these circumstances. This leads to cynicism at the working level that a new program such as Operational Excellence is just another fad of the month. People whose experience and contribution are essential to success will be reluctant to invest emotionally, become fully committed or involved, and risk the inevitable letdown if management loses interest.

At program commencement, the Executive Champion, Steering Team, and the operating leadership must all recognize that leadership discontinuities can be a serious barrier to acceptance and success. There must be agreement at the outset that Operational Excellence is the master improvement program and will remain in full force through changes in management. This must be accompanied by a solid plan and real commitment to assure any and all replacements that might take place at leadership levels are fully committed to success of the Operational Excellence program.

PROGRAM LEADER/CHAMPION

The first and one of the most important actions of the Steering Team is to appoint the Operational Excellence Program Leader/Champion. The individual appointed must be highly credible within the organization and well respected with excellent leadership, organizational, and people skills. The Program Leader must have proven ability, requisite skills, and knowledge to lead a complex improvement program, as well as energize and motivate people with widely varying backgrounds, preferences, and personalities. The individual selected must combine all this with the initiative, enthusiasm, ownership, and drive necessary to achieve success.

The Program Leader joins and becomes a fully participating member of the Steering Team and head of the Program Leadership Team. The position must be at the superintendent or equivalent level reporting to the Steering Team, with only members of the Steering Team potentially higher ranking in the organization. The Program Leader must be able, comfortable, and effective at working both up and down in the organization.

The Program Leader is a primary responsibility and full-time position, not a collateral duty. The individual must be a proven motivator and leader, with organizational stature, credibility, energy, authority, time, and ability to make things happen within the site.

The Program Leader has working-level responsibility and authority for the Operational Excellence program. These include program objectives, appointing working-level Program Leadership Team(s), fully described in Chapter 9, overseeing, facilitating and identifying, developing, and implementing improvement plans, and accountability for results.

The enterprise, facility, and Steering Team must make a commitment to have the individual selected as Program Leader remain in position at a minimum until results have been demonstrated and the first comprehensive review of the program has been accomplished. If the Program Leader is to be changed, the transition should occur during a review to give the new leader a detailed view of the program through participation in the assessment described in Chapter 17 and opportunity to participate in modifications and improvements.

Summarized Duties and Responsibilities

The Program Leader reviews the program concept and objectives with the Program Leadership Team summarized as follows and detailed in Chapter 9. The review is best conducted as a training workshop described in Chapter 11. During or after the workshop as appropriate, the program leader and Program Leadership Team refine the program mission and strategy to assure the program strategy and plan are consistent with organizational policy, strategy, and objectives and are considered doable at the working level. Any recommendations are submitted to the Steering Team for consideration and approval.

As improvement initiatives are developed by Program Leadership and Improvement Action Teams, as described in Chapter 14, and begin to solidify, the Program

Leader must assure action plans are supportive of enterprise business objectives and have credible value. Plans must also be consistent, mutually supportive, and include all the elements necessary for Steering Team approval. Resource and training requirements must be identified in the proposed improvement action plan. Any inconsistencies, questions of value, risk, or prioritization must be resolved before submission for approval.

Once improvement plans have been submitted and approved by the Steering Team, the Program Leader facilitates implementation and monitors progress, results, compliance with the strategy, interim, and final objectives. The Program Leader assures that any additional actions required to meet unexpected challenges and difficulties are identified, approved as required, and implemented. In addition to assuring interim and final results, the Program Leader must make certain actions to sustain the gains are in place and followed.

Finally, the Program Leader conducts ongoing regular reviews with the Steering Team. In some cases, a member or members of a working-level action team attaining notable results might be called on to make a brief presentation to the Steering Team. Gaining recognition by the Steering Team for exceptional results is a powerful motivating factor.

PROGRAM LEADERSHIP TEAM

At the working level, Program Leadership Team(s) are appointed to begin developing specific opportunities for improvement; refer to Chapter 9. The Program Leader facilitates the process of identifying, value prioritizing, and constructing improvement action plans in accordance with enterprise/site procedures. The Steering Team is advised of opportunities being pursued, as well as proposed actions as they develop. Steering Team feedback will be conveyed to the working-level action teams.

Returning to Figure 7.1, it should be noted that large facilities/sites with multiple operating units may have a Program Leadership Team for each operating unit depending on similarities between units.

Assistance from Finance will likely be required during the value prioritization stage to assure valuation is realistic and will be accepted when the initiative is submitted for approval. It is much easier to work with Finance to develop valuation early in the development of an initiative than to argue later after an initiative has been finalized.

> Finance is typically very receptive to working collaboratively to develop financial justification for an improvement initiative as it develops. They may even identify potential sources of value that were unrecognized by technical, professional, and operating participants. By bringing Finance into the process early, their contribution is maximized and potential resistance to an initiative sprung by surprise eliminated.

As improvement initiatives are developed and refined by the Program Leadership Team(s), the Program Leader must assure that all risks, including missing time milestones and failure of the plan to achieve results are identified, documented, addressed, and fully considered in the action plan. The Program Leader must also recognize the necessity for additional team members to provide specific knowledge or skill sets and take steps to augment teams where necessary; refer to Chapter 14. Assistance from the Steering Team may be required.

> As an improvement initiative developed, it became apparent that detailed knowledge of automation controls would be required to fully develop the improvement action plan. Accordingly, an expert with the requisite knowledge of Programmable Logic Controllers (PLC) was added to the team, assuring all knowledge for a complete improvement plan was resident on the team and available.

WHAT YOU SHOULD TAKE AWAY

To gain full success from Operational Excellence, it is imperative to have a fully defined and active leadership structure. Beginning with an Executive Champion, extending through the Steering Team, Program Leader and Program Leadership Teams all must be top-caliber people with full ownership, active involvement, and commitment to the program. If promotions result in changes to the leadership team, replacements must have equal ownership and commitment to the program to assure continuity and continuing success.

8

SAFETY AND HUMAN PERFORMANCE EXCELLENCE

PETER G. MARTIN
Vice President Invensys

As mentioned in Chapter 3 and illustrated in Figure 3.1, safety, health, and environment (SHE) and human performance excellence are essential elements of Operational Excellence that contribute to and cross all functions. SHE and human performance excellence are mandatory to gain the working culture, commitment, and ownership required for successful Operational Excellence.

SAFETY PERFORMANCE EXCELLENCE

SHE performance excellence is a constant requirement across all domains of Operational Excellence. In the context of Operational Excellence, safety encompasses the safety and health of people, systems, equipment, products, plant, and the environment. It includes process and individual safety. Figure 8.1 shows the relationship between safety and the real-time profitability variables. Safety, heavily considering risk, often provides a primary constraint in industrial operations with processes and materials that can pose safety risks and require active precautions.

Risk Management

Risk management, explored more fully in Chapter 12, is a critical aspect of Operational Excellence. Industrial organizations tend to be very conservative with respect to process and individual safety and minimize risks as much as possible. Safety risk tends to vary with the operating process, local environment, intensity

Operational Excellence: Journey to Creating Sustainable Value, First Edition. John S. Mitchell.

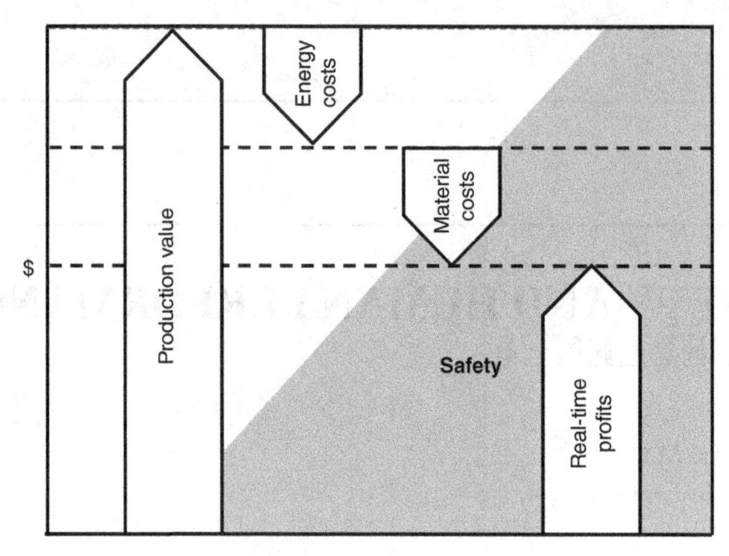

Figure 8.1 Safety impact on real-time profitability.

of operation, equipment maintenance, utilization requirements, and products. Thus, safety and risks associated with safety excellence often add an additional constraint on profitability.

> In operations with huge costs of downtime, sometimes approaching millions of dollars per day, there is often a corresponding tendency to make decisions on the basis of a perceived necessity to restore operation without fully considering risk. Several incidents have illustrated that decisions made in haste to save millions in potential losses actually resulted in billions of real losses in addition to severe damage to enterprise reputation and credibility.

A current safety risk is seldom, if ever measured in industrial operations. Therefore, most industrial operations set their safety risk at worst-case levels on an ongoing basis, which tends to constrain profitability more than necessary. The result is that business managers in the operation may consider safety as something that is confining and needs to be dealt with rather than an integral component of the profitability of the operation. Consequently, almost all industrial organizations include a SHE function as a governing support group to the operation.

Developing a safety performance excellence strategy requires that the current safety risk within each section of the operating enterprise is effectively measured on an ongoing basis. Understanding the actual current safety risk might enable the plant operations personnel to push throughput higher during lower risk periods, which could result in significant improvements to profitability. It can also help operators avoid unnecessary risk brought on by unsafe events. The principles of risk analysis and Failure Modes and Effects Analysis (FMEA) described more fully in Chapter 12 provide guidance.

Developing a safety risk performance excellence strategy must begin with an extensive design audit. Particularly, in an older facility where SHE and process safety requirements may have increased and become more stringent since commissioning. Are all of the latest safety-related requirements for systems, assets, equipment, and people in place, up-to-date, and observed?

After a major incident, similar facilities were directed to conduct a detailed safety audit. Findings indicated that over the years, projects to increase production had not been accompanied with a thorough safety analysis. In many cases, protective systems were undersized for the uprated conditions. Results of the analysis forced many facilities to reduce production rates until safety systems could be re-engineered and updated for the new conditions.

The conundrum in implementing a safety performance excellence approach is in being able to effectively measure real-time safety risk. Although a perfect safety risk measurement system is still far from developed, industry can certainly provide reasonable, although typically still conservative, real-time safety risk measures by partitioning safety risk measurement into two fundamental components.

Operational Safety Risk The first component is operational safety risk. This is a risk assessment based on the faithful and timely execution of human-based processes. These include statutory, regulatory, and asset integrity tests and inspections, hazardous operations (HAZOP), safety equipment inspections, Occupational Safety and Health Administration (OSHA) training and inspections, preventive and predictive maintenance processes, and similar defined activities intended to reduce operational safety risk. Where applicable, Safety Integrity Levels (SILs) need to be validated along with risk mitigating Safety Instrumented System (SIS) and other requirements. If all of the activities and processes are accomplished according to the prescribed requirements and schedules and all remediation activities are faithfully undertaken, safety risk will be reduced. Since this is not a precise safety risk assessment, it is appropriate to measure operational safety risk on a fairly simplistic basis—high, moderate, or low risk. The operational safety risk can be conveyed to plant personnel through a simple green, yellow, red dashboard.

Conditional Safety Risk The second aspect of safety risk is conditional safety risk based on the identification of actual operating conditions in the production operation. Even when all the operational safety activities are rigorously undertaken, history has demonstrated that events or sequences of events can occur that could lead to an SHE incident. Safety incidents often occur during a chain of unexpected anomalies. Decisions are typically based on the best information available. Added risk incurred by a cascade of decisions made during periods of abnormal or upset operations is not always considered.

Serious incidents have occurred where multiple barriers in place to reduce probability are compromised one by one for expediency without recognizing the added risk incurred at each step.

Potential operational risk can be identified through an asset integrity process. Lead indicators of safety incidents can be identified by reviewing the historical record of the operation as well as those of other similar operations. Procedures are then developed to counter a potential incident and minimize risk of recurrence. With training and reminders, people in decision-making positions will have awareness of how risk must be factored into decisions that may be made under intense operating conditions. Risk is reduced by recognizing and responding to probable potential events in a predetermined manner before they fully develop.

Fortunately, due to the high level of automation achieved in many industrial operations over the past three decades, most have rich historical databases. These databases can be analyzed for the points in the past in which safety incidents or near misses happened to isolate potential lead indicators that may have led into the events as well as elements of risk. Building automated workflows that are executed whenever lead indicators of potential safety variations arise in the operation provides warning of potential risks in time to react before they become actual problems and thereby lower risk.

Until a complete historical database of an operation over an extended time frame is developed, any potential incident analysis may be incomplete. Thus, it is important not to be too liberal with the safety risk setting derived from conditional safety risk analysis. As with operational safety risk, it is recommended that a fairly simplistic, green, yellow, red indication dashboard be used to communicate the real-time conditional safety risk. For highly dangerous conditional safety risk issues, SISs must be installed to automatically and immediately respond to an impending problem.

Safety Risk Dashboard Since having multiple safety risk indicators may prove confusing to plant operations personnel, it is recommended that a composite safety risk dashboard be developed, Figure 8.2. In the spirit of being conservative with respect to safety issues, the composite safety risk measure should display the highest safety risk from the operational and conditional safety risk dashboards. With the composite safety risk indicator as part of the performance dashboard, operators will know when they are able to safely push the performance of the operation and increase profitability. The dashboard should have "drill down" capability so that operators can identify and assess the precise area that is governing risk. The result will be increased profits within a safe operational environment. This is the primary objective of a safety performance excellence strategy. Implementing safety performance excellence serves to pull safety right into the mainstream business processes by directly associating safety with profitability. From an organizational perspective, the SHE function typically becomes much more mainstream within the industrial operation.

HUMAN PERFORMANCE EXCELLENCE

As stated in Chapter 6, human performance excellence must be attained across the entire enterprise and spectrum of Operational Excellence. The idea of maximizing the performance of personnel in industrial operations is obvious and has been around for

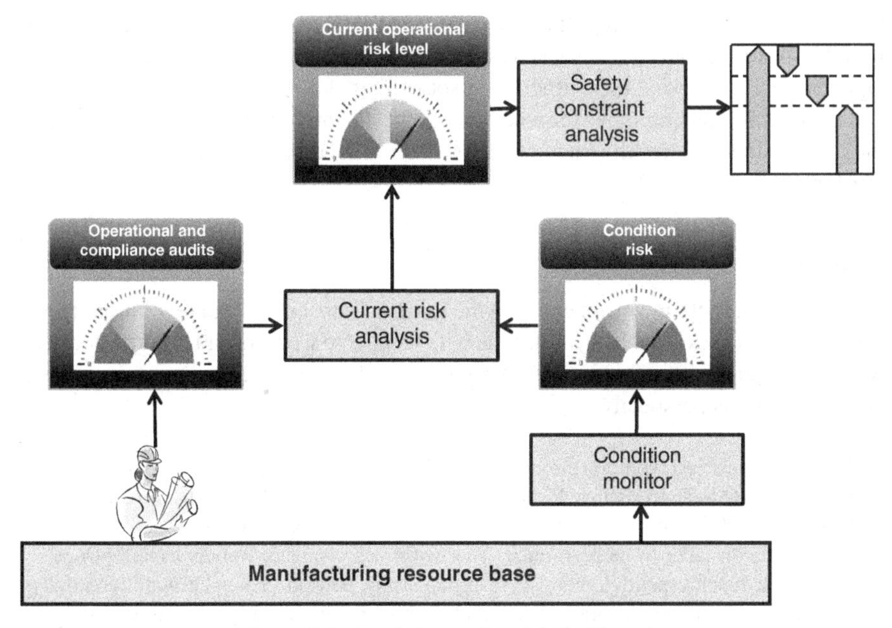

Figure 8.2 Real-time safety risk dashboard.

a long time. In the initial stages of the industrial revolution, industry was challenged on how to effectively use an unskilled and uneducated labor force to make industrial operations work efficiently. Frederick Taylor proposed a solution to this dilemma by introducing the concept of "scientific management" to industry. In essence, scientific management was based on defining a small set of repetitive functions that each laborer would perform and never allowing them to deviate. This essentially turned the labor force into a collection of nonthinking production line robots.

> Trained, committed people with ownership for success are the most valuable asset within the enterprise. A strategic training and human development plan that provides the essential competencies to meet the program and overall business/mission strategies and objectives and drives results is imperative.

Automation Technologies

Plant engineers worked to minimize the negative potential impact that such an unskilled labor force might have on the operation by either replacing operators with automation technologies or applying technology to accomplish the more advanced functions, as well as to limit the responses operators are allowed to make. To support this, control systems were, and still are, designed to interact with operators on the basis of the philosophy of "operation by exception." This means that the operators are chartered with doing only limited and well-defined functions until an exception or alarm condition occurs, then they are to handle the exception condition and go back to what they were doing. It is not surprising that one of the most prevalent items found in control rooms are newspapers, motorcycle, hunting, and fishing magazines! Operators have to spend their time doing something. The point is that as automation and information technologies have advanced significantly over the past few decades, the attitude that the frontline personnel are merely laborers has really not advanced much at all.

Improvement Culture

Some academics claim we are entering the third or fourth industrial revolution. However, when it comes to industry's attitude with respect to workers and tasks performed, we are still stuck in the first industrial revolution. Operational Excellence demands a change from this perspective.

> From the industrial revolution through "scientific management," and until recently, organizations were built on control. Workers were instructed what, how, when, and expected to follow to the letter. Many, if not most labor contracts, were written with detailed restrictions on tasks to be performed. The world has changed. Robots have replaced humans in much repetitive work. For an increasing portion of human work, control has become a limitation; brainpower is all important. Performing tasks more safely and effectively at greater speed is all important. Although there must be some level of control, it must be balanced to gain greatest effectiveness. For managers, contribution, results, and value delivered must be the measures of performance through all levels of an enterprise, not how many people are under the individual's control. Align people, optimize processes, and accelerate toward Operational Excellence.

Today, frontline personnel are both educated and experienced at levels never previously attained. Most are highly computer literate; all will be in a few short years. It is time to use the improvement culture demanded by Operational Excellence to change the culture of industrial operations, so the frontline employees assume the role of performance managers committed to excellence and continuous improvement rather than rote laborers. These employees are an expensive and talented resource base that has been grossly underutilized. Human performance excellence is based on changing the attitude toward valuable human resources, empowering them with the information they need to make value generating decisions, and holding them accountable for results and rewarding success.

Performance empowerment involves providing the correct decision support information and guidance to each employee in the operation on the time frame required to make performance improving decisions—in real time. The real-time profitability measurements discussed in Chapter 6 coupled with real-time Key Performance Indicators (KPIs) serve as the basis for empowerment. The challenge to resolve is providing all the information needed for accurate decisions while preventing personnel from being overwhelmed with overload from less important data. As mentioned in Chapter 6, experienced people with the equivalent of a high school education are able to learn how to balance variables similar to profitability and gain optimal performance in a few weeks.

The situation in maintenance is much the same. Maintenance technicians under any name are typically directed to restoration. Operational Excellence emphasizes ownership for performance and reliability and demands efforts to identify and understand why a problem occurred and to improve conditions so the problem does not reoccur.

A process is therefore required to focus each person on the exact performance measures required for excellence in their job functions. This can be performed by decomposing the current manufacturing strategy to each functional domain and setting up the appropriate real-time dashboards and scorecards, as well as real-time workflow guidance. The objective is to provide each person with the exact information they require to make good decisions that contribute to business/mission objectives and value. Interestingly, these people are already making hundreds of decisions a day without any true understanding of how these decisions improve or diminish value. By merely providing an understanding of the value of their decisions, operations and maintenance personnel tend to make much better value generating decisions. The result is continuously improving and sustainable profitability. This is the essence of human performance excellence.

WHAT YOU SHOULD TAKE AWAY

Safety (SHE) and human performance excellence are essential elements of Operational Excellence that contribute to and cross all functional domains. SHE and human performance excellence, the working culture, commitment, and ownership are required to achieve successful Operational Excellence.

Business-focused Operational Excellence strategies may initially appear to present daunting challenges to industrial operations. Coordinating efforts within Operational Excellence, partitioning the overall strategy into areas of individual control, and demanding safety, human, and work performance excellence will generate the required results.

Minimizing SHE and operational risk requires identifying potential sources of risk and developing safe and solid procedures that can be applied when precursors to a potential incident are identified. Everyone must be trained to fully consider the risk during any abnormal event requiring decisions under pressures of time.

Taken together, safety, human, and operating excellence provide the results necessary for world-class performance.

9

DEFINE THE PROGRAM AND PROGRAM OBJECTIVES

The initial, Define stage of an Operational Excellence program is one element that sets it apart from other improvement programs. Unlike many other improvement programs that are essentially fixed in process and implementation, only needing to define an opportunity (e.g., Six Sigma), an Operational Excellence program itself must be defined for enterprise- and site-specific mission/business objectives, strategy, and conditions. That is, where are we now, where do we need to be, and what will get us there?

The top, enterprise business/mission objectives and strategy are normally established above the operating level. Grow the business and improve profitability/mission effectiveness are examples of top-level strategic objectives. Although growth objectives may extend to the operating level, improving profitability/mission effectiveness is the most likely operating level objective. For this, we need to define exactly where we are, what we are building on, and what improvements can be made to safely add greatest sustainable value contribution in the least amount of time. Other questions need to be answered about the organization and administrative system: are improvements necessary to attain the objectives, considering the current system and the practicality of major change? Will the improvement program be implemented enterprise wide or as a proof of concept pilot: unit, site/facility?

Operational Excellence, including the program itself, demands constant improvement. All elements must be reviewed at regular intervals, improved, and even redefined. This is to ensure that the program remains evergreen and is able to meet all internal and external requirements and conditions, which themselves are likely changing, especially as improvements take hold.

Operational Excellence: Journey to Creating Sustainable Value, First Edition. John S. Mitchell.

At the beginning of an improvement program, one site established an annual loss of $1.5 million as a minimum threshold for considering improvement actions. Within a few years, the high value opportunities for improvement had been eliminated, leading to a reduction in the minimum value threshold to $750,000.

EXECUTIVE LEADERSHIP

As outlined in Chapter 7, a clear, compelling executive vision is the essential starting point for Operational Excellence. The vision defines the end state, where the enterprise must be with measurable objectives including time to get there. An accompanying enterprise strategy identifies in general terms how the objectives will be accomplished. A process flow diagram is illustrated in Figure 9.1.

To establish and maintain close support for the Operational Excellence program, the Executive Champion might consider creating a position in the organization such as Vice President or Director of Operational Excellence. As detailed in Chapter 7, a staff person may be appointed as a fully empowered deputy to actively promote and facilitate the Operational Excellence program on a day-to-day basis. In either case, the deputy must be seen by all in the organization as the direct representative of the Executive Champion with real power to assure a successful outcome. The executive champion or executive deputy must demonstrate commitment through personal, active involvement with the Steering and Leadership Teams during the early stages of the initiative. The oversight continues through maturity. The sponsoring executive's deputy will need to clearly and authoritatively answer the questions "what exactly is required, and when must it be achieved?" Failure to do so is an adverse statement reflecting on the essential support required from the executive champion.

Regardless of delegating direct participation, executive leadership is all important to the success of Operational Excellence. The executive champion must stay engaged, actively driving the program with regular communications, knowledgeable encouragement, and personal interaction during site/facility visits. Continuing executive commitment, drive, and support is an absolutely essential ingredient for success. Active, visible commitment from the top will flow through the organization to gain commitment, support, ownership, and responsibility at the working level.

From here on in this text, the Executive Champion and executive deputy, if appointed, will be considered one and the same.

OPERATING ORGANIZATION

Steering Team

Appointing the right Steering Team is crucial to success; refer to Chapter 7. The team must be assembled with strong, enthusiastic, and supportive individuals at high levels within the operation who possess real power to get things done.

The program Steering Team is responsible for oversight and establishment of the organizational environment that will assure success. The Steering Team must

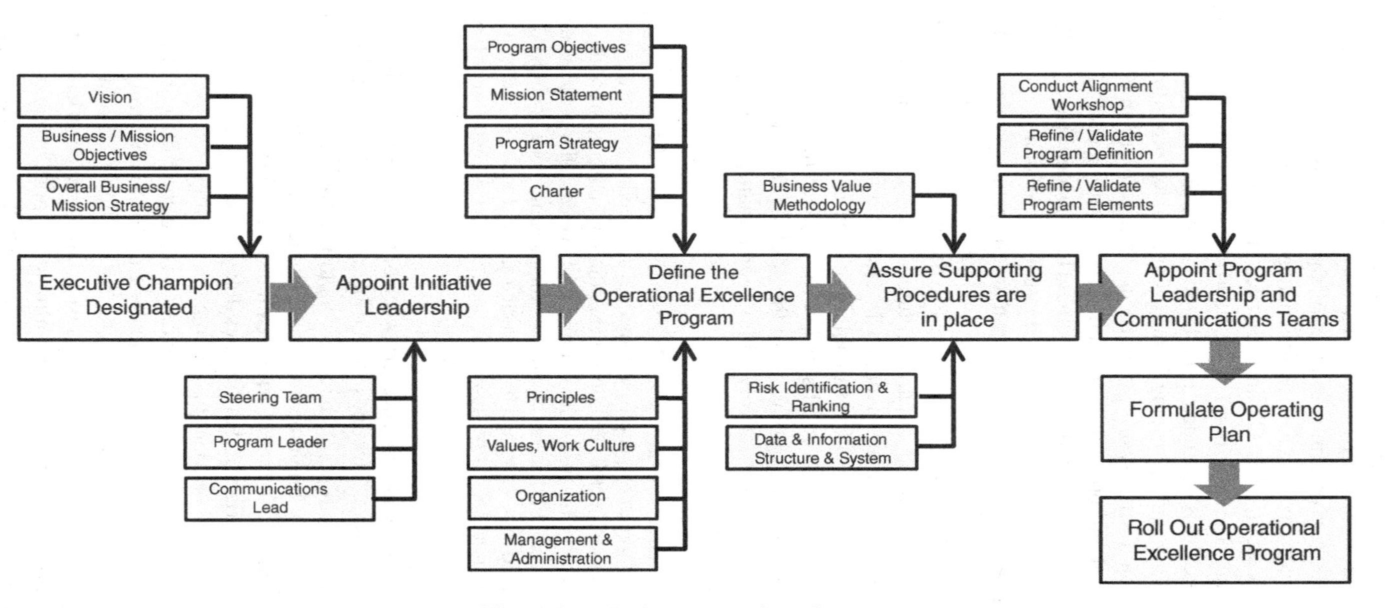

Figure 9.1 Define process flow diagram.

be strategically oriented and have authority within the organization to resolve any functional and departmental differences that may inhibit full exploitation of the Operational Excellence program. Members of the Steering Team should include the site/plant manager (ex officio), senior managers from operations/production, maintenance, engineering, HR, IT, and finance, and the Operational Excellence program leader. The Steering Team is responsible for defining the Operational Excellence program. This includes establishing program objectives, principles, and strategy, reviewing and approving improvement plans, providing guidance for implementation, monitoring results, facilitating interdepartmental coordination, removing institutional and procedural barriers, and rewarding success.

The most senior and committed member should be considered to chair the Steering Team. With all equal, a senior manager from operations/production is a logical choice.

Strategic direction and program facilitation are the primary roles of the Steering Team. It is crucial for all members to recognize that the Steering Team is a strategic body that advises, recommends, and reviews. The Steering Team must concentrate on program definition, organization and architecture, objectives, and results. The Steering Team must avoid detailed involvement in tactical and implementation issues, except where experience may offer valuable guidance or where organizational or functional barriers may impede results and success. With this stated, the Steering Team must be aware of what is going on in the working-level improvement teams. They must act promptly and proactively without micromanaging the process if problems occur or the teams encounter barriers.

A Steering Team meeting at a large, complex operating facility became bogged down in detail. There were too many people participating, the meeting attempted to cover too many issues in too much detail, and, in general, did not appear to have the strategic or value focus necessary for discussions at this level. Three major presentations were delivered. Because only 1 hour was allotted for the meeting, none could be discussed in any depth. Detailed reviews should be conducted at the Leadership Team level. "Dog and pony" presentations should be limited to one per meeting, with strict time limits that allow discussion.

The Steering Team typically meets every week or two at the beginning of the Operational Excellence program. Meetings should be used to drive toward performance objectives and identify weaknesses and barriers to resolve. Meetings should not be used for detailed reviews of implementing tactics, unless there are specific challenges that can only be reconciled at the Steering Team level. Performance, deviations from required performance, barriers and difficulties delaying results, and potential requirements for action should all be identified before the meeting.

The words stated earlier regarding continuing visible commitment, drive, and support from the executive champion hold equally true for the Steering Team and are an absolutely essential ingredient for success. In addition to leading a site/facility improvement program, the Steering Team must be an active cheerleader for success. Here again, it is impossible to gain the commitment, support, ownership, and responsibility at the working level necessary to create maximum value if people don't see an active, visible equivalent from management, in this case the Steering Team.

Appointing initial operating leadership outlined in the following sections as well as developing an initial program strategy should be completed in no more than 1 month from program initiation. In addition, the Steering Team should endeavor to develop and release a site-wide communication conveying program objectives, necessity, and benefits close to program initiation. This will be followed by status, successes, and results at regular intervals, approximately monthly. Communications are directed to reinforce the initiative and address issues and concerns. Communications will be addressed in greater depth in Chapter 10.

Program Leader/Champion

The first task of the Steering Team is to appoint the Operational Excellence program leader/champion who will join and become a fully participating member of the Steering Team. The program leader is crucial. He/she will be responsible for energizing and driving the program and personally accountable for results. As stated in Chapter 7, the individual selected must be highly respected and qualified, with proven ability to lead a major effort and people with widely varying backgrounds, preferences, and personalities. The program leader must be at an equivalent level within the organization of operations/production or maintenance superintendent and report directly to the site/facility manager and Steering Team. The program leader must be able, comfortable, and effective at working both up and down in the organization.

The program leader is a primary responsibility and full-time position, not a collateral duty. The individual must be a proven motivator and leader with organizational stature, credibility, energy, authority, time, and ability to make things happen within the site.

The facility and Steering Team must make a commitment to have the person selected remain in position at a minimum until results have been demonstrated and the first comprehensive review of the program has been accomplished. If the program leader/champion is to be changed, the transition should occur during a review to give the new leader a detailed view of the program through participation in the assessment, defined in Chapter 17, and opportunity to participate in modifications and improvements.

PROGRAM PLAN

With the Steering Team formed and program leader selected, the real task begins—defining the Operational Excellence program. Depending on familiarity with requirements and the Operational Excellence program itself, it may be advisable to initiate the definition phase with an alignment workshop similar to what will be described in Chapter 11. The alignment workshop is designed to assure that the entire Steering Team is of one mind regarding enterprise business/mission requirements and objectives; understand how the Operational Excellence program works and contributes to improving business/mission performance and results. In addition, everyone in the Steering Team must be in agreement with program definition and objectives necessary to fulfill enterprise requirements, program principles, strategy, values, and a preliminary organizational structure to achieve objective results. The preliminary plan developed by the Steering Team will be reviewed by the Program Leadership Team formed later in the process. Any recommendations for modifications are discussed with the Steering Team for approval and finalization as the program plan.

Program Objectives

Program objectives: specific results, including time, to meet the visionary future state expressed in the enterprise business/mission objectives.

> "It's one thing to establish objectives. It's quite another to get people to believe passionately they absolutely must reach them to succeed: take ownership and make a full commitment to objective results!"
>
> Plant Manager

The Steering Team translates enterprise objectives into specific objectives for the Operational Excellence program. Since enterprise objectives are typically mission/business directed, the Steering Team accomplishes the initial translation into the operating domain. This includes establishing the connection between enterprise business/mission success and operating variables to be addressed by the Operational Excellence program. One typical enterprise business objective might be to reduce costs per unit production; refer to Chapter 5. Depending on business conditions, this could translate into some combination of producing more at the same cost through increased availability, production rate, quality, or improving production/mission efficiency at the same output to reduce operating costs. As will be noted later, this initial translation is guidance; the issue will be addressed in detail and refined by Program Leadership Team(s).

> A company that could sell everything produced without any price erosion concluded that concentrating on increasing throughput while holding fixed operating costs constant or nearly so was the most effective way to improve. This led them to an initial focus on production rate and quality.

From program objectives, the program mission, principles, overall strategy, and plan will emerge.

Mission Statement

Mission statement: an overall statement of measurable objectives that must be accomplished to attain the future state.

The initial mission statement for the Operational Excellence program will be written by the Steering Team. The mission statement defines the program purpose and objectives as succinctly as possible. It is designed to create positive awareness, overall guidance, direction, motivation, and ownership to participants and nonparticipants alike. The mission statement should include measurable objectives and tangible benefits so that everyone is aware of the destination and what success will mean to them.

A typical mission statement might read as follows:

> To assure continuing success and fully support all current and long-term business and operational objectives, the Operational Excellence program will safely establish and sustain an improvement of at least 20% overall operational effectiveness within 5 years measured by improved OEE and reduced operating cost. The program will improve performance and value delivered, build ownership for excellence and improvement, and assure job security.

To go along with the mission statement, the Steering Team should suggest progress (activity) and results metrics (see Chapter 15 for definitions) that are as specific as possible along with the means of measurement. The Value Model proposed in Chapter 5 can be the basis for linking operational effectiveness to business/mission value objectives.

The Mission Statement will be reviewed, potentially refined, and improved later in the program initiation phase and during periodic reviews by the Program Leadership Team. Any proposed alterations or improvements will be submitted to the Steering Team for final approval.

Charter

The program charter is a high level statement of the objectives, overall principles, organizational values, requirements, and strategy for the Operational Excellence initiative. It is derived from and states the enterprise's position, intentions, and objectives that drive the program. The charter establishes the commitment to satisfy requirements for SHE, mission and performance effectiveness, continuous improvement, and sustainability. The charter initiates and governs the Operational Excellence program and translates enterprise mission/business objectives into program objectives. The charter includes the Operational Excellence program mission statement and details the program strategy. It describes in general terms the means to fulfill the strategy as well as the framework and guidelines for the program, including organization, administration, and controls. It is imperative for the charter to be in compliance with and fulfill all requirements of the ISO55000 series.

The charter identifies potential sources of value and how value will be calculated to satisfy enterprise mission/business objectives. The enterprise process for determining risk is identified within the charter, as well as the method for ranking and control if not included in the risk analysis procedure. A procedure for recording, analyzing, and eliminating incidents and failures such as Root Cause Analysis (RCA) should be identified in the charter along with any amplifying instructions for use. It is important to refer to enterprise/site procedures and instructions rather than quoting directly. The former allows revisions to a procedure without the necessity of revising derivative documentation.

Principles

The foundation principles of Operational Excellence were introduced in Chapter 3. The Steering Team should develop a series of specific facility/site, mission/business principles to assure greatest understanding of program objectives and benefits. Principles must also include the individual and collective responsibilities that are necessary to guide program implementation. Principles are often written very personally beginning with "We will."

When stated correctly, program principles will:

- Improve organizational awareness of mission/business imperatives and the activities and actions necessary to safely maximize mission/business success
- Assure integrity, honesty, and mutual trust in all activities and relationships, including employees, suppliers, and outside interested parties
- Increase the energy, initiative, ownership, responsibility, accountability, teamwork, and effectiveness of the entire organization
- Increase the organizational motivation and commitment to excellence, continuous improvement, and sustainability in all activities
- Shift greater control, ownership, and accountability to the working levels of the organization and promote behaviors that produce greater individual initiative and responsibility
- Energize personal commitment, positive contribution; improve relationships, communications, teamwork, and response to problems
- Develop total commitment to excellence and continuing identification of additional opportunities for improvement
- Reduce non-value-added activities, unnecessary bureaucracy, excessive layers of management, and duplicative efforts.

Values

Values establish what the Operational Program stands for. Values establish behavior, behavior defines the working culture, and the working culture drives results. Thus, a forceful exposition of values: what we stand for, is absolutely essential at the beginning of the Operational Excellence program.

> A strong effort to achieve optimum values and working culture at the beginning of the Operational Excellence initiative greatly facilitates gaining the necessary improvements in process and procedure. Essential improvements in process, procedure, and habits will flow naturally if the correct organizational values and work culture are in place to build on. Attempting to improve processes and/or procedures while leaving organizational values and culture inconsistent with commitment and ownership for excellence to be addressed later in the process allows resistance to persist and significant barriers restricting success to remain in place—the initiative is likely to fail.

The Steering Team establishes an initial list of values that are reviewed by the Leadership Team. Recommended modifications, changes, and additions are submitted to the Steering Team for approval. Values that must be expressed and continually demonstrated include the following:

- Total commitment to safe, responsible operation, enterprise business/mission objectives, and creation of maximum sustainable value
- Honesty and integrity in all elements of the enterprise, the operation and Operational Excellence program, and personal relationships
- Mutual trust, respect for the individual, honest, open, fair dealings with all employees, transparency, and promises kept
- Positive attitude and constructive relationships throughout
- Curiosity, initiative, ownership, support, and accountability encouraged at all levels
- Quick to identify and correct mistakes: mistakes initiate learning and improvement not blame
- Decision responsibility and empowerment for action delegated to the working levels whenever possible commensurate with risk
- High degree of teamwork, continuous coaching for mutual success
- Frequent, open communications
- Compensation for contribution to success and an effective working culture.

In an employee survey conducted at the beginning of an improvement initiative, employees were asked to rank, by importance, organizational characteristics that were most important for their job satisfaction. Mutual trust, respect for the individual, and honest, open, fair dealings ranked at the top. Compensation ranked near the bottom. It should be noted that most employees were long time at the facility and likely paid well for the geographic area.

Program Strategy

Strategy: *description of major actions and sequence of implementation necessary to meet objectives—attain a future state. See Chapter 13 for an example of a military strategy.*

The program strategy is the overall design for optimally achieving Operational Excellence. It is based on, derived from, and fully supportive of the enterprise strategy and business/mission objectives. The program strategy builds from the mission statement and defines overall requirements to meet program objectives; what to do: for example, increase mission compliance, production throughput, and reduce operating costs. The strategy should state the order of importance and priority of requirements.

Developing the Operational Excellence program strategy typically begins with the mission statement. Full compliance to SHE, statutory, legal, and code requirements are stated. The strategy converts principles into overall actions and activities to gain greatest sustainable lifetime value and return. It is constructed to assure that activities are aligned and sequenced to work together with maximum efficiency and consistency and contribute demonstrable value.

The program strategy forms the basis for investments, allocating resources including people, processes, and technology to optimally achieve business, mission, and program objectives. A well-thought-out and logically constructed strategy assures the Operational Excellence program is directed by lifetime value and effectiveness principles rather than short-term cost considerations.

The program strategy must be clear, universally understood, communicated, and readily available to all. It is reviewed and updated periodically in order to assure that the strategy remains effective, consistent, and in full alignment with the organizational strategic plan, program charter, and other organizational policies and strategies.

Developing and executing an optimum Operational Excellence strategy can be the difference between excellence and mediocrity. An effective Operational Excellence strategy aligns and optimizes objectives with available people, processes, and technology resources. Success demands strong results metrics, as described in Chapter 15, which align to strategic objectives, track performance, and provide the visibility necessary for sustaining results and continuous improvement.

Organization: Management, Control, and Administrative Systems

The program charter must include a clear, unambiguous program organization and management system. Figure 7.1, in Chapter 7, illustrates a typical Operational Excellence organization. The organization provides the structure to meet business/mission and program objectives while gaining maximum ownership, contribution, and effectiveness from all of the people involved. The ISO55000 series has a number of specific requirements for asset management organization, control, management, and administrative systems. Since asset management will be included within Operational Excellence, the program organization, management, and administration system must be in full compliance with ISO55000.

> "Excellent companies are made up of excellent teams made up of excellent people."
>
> Wayne Vaughn, Harley Davidson

There are several prerequisites for establishing the Operational Excellence program organization. First and foremost, the very best, most highly motivated and dedicated people must be assigned to the program. Ownership and a real commitment to success are essentials. There must be a concurrent commitment to maintaining stable leadership throughout the formative period. Any replacements within leadership must be fully committed to the Operational Excellence program and its success.

Established by the Steering Team, the organization includes scope, applicability, and boundaries of the program. It thoroughly defines roles and responsibilities in a RASCI (Responsible, Accountable, Support, Consult, Inform) or equivalent form supplemented with explanatory text. Figure 7.2 illustrated one example; Figure 9.2

Root Cause Analysis (RCA) Process	RCA Program Lead	Analysis Lead	Operations Superintendent	Maintenance superintendent	Engineering Manager	SHE Manager	Site / Plant Manager	Support Personnel (SME's)	Personnel Involved
Establish and maintain RCA procedure and training, assure skills	A	I	C	C	C	C	C	I	
Convey program necessity and benefits to site personnel	A	I	I	I	I	I	I	I	
Determine criteria and type of analysis required	A	C	C	C	C	C	C	C	
Preserve and collect data	S	R	S	S	S	S	S	S	A
Assemble analysis team	S	A	S	S	S	S	S	S	
Conduct the investigation / analysis	R	A	S	S	S	S	S	S	S
Identify remedial actions	R	A	C	C	C	C	C	C	
Identify potential similar failures	R	A	S	S	S	S	S	S	
Implement remedial actions to prevent similar incidents	A	R	S	S	S	S	S	S	
Document analysis results and remedial actions taken	S	A	I	I	I	I	I	I	I
Publicize findings to include remedial actions	A	R	I	I	I	I	I	I	I
Monitor effectiveness	A	C	I	I	I	I	I	I	I

Figure 9.2 Root Cause Analysis (RCA) Responsible, Accountable, Support, Consult, Inform (RASCI) diagram.

illustrating a program component, in this case RCA, is another. Roles and responsibilities must establish clear authority for decisions considering potential risks. As guidance, working-level people should be empowered with maximum authority to decide how to most efficiently perform activities commensurate with SHE compliance and risk.

Organizational Objectives

Constructing the most effective organizational structure for the tasks at hand is an essential within Operational Excellence. The following are key objectives of the Operational Excellence organizational structure that will be amplified in this and the following chapters:

- Flatter organization
- Multifunction teams at all levels in the operation
- Empowered decision makers: everyone obligated to support decision, decision maker obligated to review, adjust and revise as necessary
- Greater flexibility in work processes.

Clarity of leadership, effective organization, engagement, commitment, and ownership of the work force, and collaboration between departments and functions are essentials to create success.

The Operational Excellence organization must consider the existing site/facility organization—particularly those parts of the organization that will interface with program activities. Production, maintenance, safety, finance, quality management, IT, and HR are examples. All should be referenced in the organizational charter with a statement that the program will be aligned with and in full compliance with current procedures, roles, responsibilities, authority, and decision processes. Any exceptions, areas of potential conflict between the Operational Excellence program and established administrative systems, processes, and procedures, will be identified and referred to the Steering Team by the Program Leader with recommendations for resolution. In all cases, the Operational Excellence program roles and responsibilities must be configured to facilitate interchange, dialog, and cooperation across functional divisions.

In terms of organization and selecting people to occupy positions within Operational Excellence, it is most important to select the best, most willing and capable of working in, and energizing a team-based, multidiscipline, multifunction process. Improvement processes have been observed to thrive when directed by production and maintenance managers and maintenance and reliability engineers. In every case, it is the individual who is the key ingredient. Selecting the correct people for leadership is an essential and likely the single most important element to achieving success with Operational Excellence.

The Operational Excellence program organization should be widely communicated within the organization. Questions and comments are actively encouraged. Roles and responsibilities should be reviewed as part of the overall program review process, with modifications and improvements made as indicated. The success of the organization and administrative system is reported periodically to the program executive champion.

Business and Risk Models

As stated earlier, the Steering Team will assure that enterprise-approved business value, risk assessment, and failure/incident analysis procedures and models are in place, in conformance with appropriate standards. All will be designated as fully applicable to Operational Excellence and will be up to date. The business value model is the origin of definitive program metrics that will demonstrate contribution to mission/business objectives. A risk assessment procedure is essential to identify and accurately prioritize problems with potentially serious consequences that could, but may not yet, have occurred.

If business value, risk models, and failure/incident analysis procedures are not available, complete, or accurate, the Steering Team will assure the omission is corrected as quickly as possible. These procedures are the basis for the value prioritization that is an essential of Operational Excellence improvement initiatives.

FORMULATE PROGRAM OPERATING PLAN

With objectives and strategy in place, the Steering Team expands the program charter into the outline, contents, and initial draft of the first sections of a program plan as the basis for the next, planning phase of the program. The program plan, typically covering 5 years, links corporate business objectives to operating goals and defines the actions and results necessary to meet objectives. The completed program plan, fully defined in Chapter 14, describes principles, actions, and activities to meet objectives, including resources required and measures of performance, as well as time required and probability of success. It includes requirements for sustainability and continuous improvement of the Operational Excellence program. The plan is a living document that is reviewed and revised at regular intervals to account for changing conditions and circumstances.

> There are many advantages to a formal program plan and the planning process that compile mission, program objectives, strategy, and individual improvement action plans into an integrated, descriptive document. The program plan is the link between corporate business/mission objectives and the operating goals, tactics, and improvement initiatives necessary to achieve the required results. The process of developing the plan ensures that objectives, principles, strategy, improvement initiatives, and measures of performance are fully defined and understood, prioritized, and communicated throughout the organization.

An Operational Excellence program plan should follow a standard format developed by the Steering Team. It should include all the information necessary to define the specific objectives and results required to demonstrate conformance to enterprise business/mission objectives. Within a 5-year program plan, the first 2 years are generally described in great detail. The plan becomes progressively less detailed in years three through five concentrating on overall expectations, objectives, and broad ideas for fulfillment.

The plan details the program principles and strategy, describes the rationale and validity of each improvement initiative, probable contribution to objectives, and identifies potential challenges and barriers that may require resolution. It defines requirements for risk assessment, management, and control and includes a commitment to continuous improvement and sustainability. Specific objectives, tasks, and metrics (KPIs) must be quantified as much as possible. Improvement action plans must provide activity and results metrics, as mentioned in Chapter 16, to judge whether objectives were achieved and, if not, why not.

As stated, ISO55000 series, Asset Management, governs one of the major functional areas included within Operational Excellence. Thus, the Operational Excellence program and plan must conform to and be fully compliant with all requirements of ISO55000. Specifically these are strategy, objectives, the administration and control systems, and the audit, assessment process to be described in Chapter 17.

The Operational Excellence program plan is optimally constructed in two phases. The first phase through organization is developed by the Steering Team and later refined by the Program Leadership Team. The remainder will be assembled by the Program Leader and Leadership Team, as Improvement Action Teams define specific improvement plans to fulfill program objectives and strategy. The plan will be approved by the Steering Team.

A suggested format for the program plan follows. The plan will be revisited in Chapter 14.

- Enterprise vision, overall mission, and business/mission objectives and strategy (from charter)
- Program mission and mission statement to meet enterprise business/mission objectives (from charter)
- Executive summary (normally written after the basic plan is complete)
  - Strategy to meet enterprise mission/business objectives
  - Overall results expected: value contribution to enterprise business/mission objectives, time and resource requirements, potential challenges, and probability of success
  - Summarized improvement action plans (added as plans are approved)
    - — Rationale and opportunity for improvement
    - — Results and future state including KPIs
    - — Value delivered
    - — Major actions and activities to gain objective results

    — Resources required
    — Risk, including potential barriers, challenges, and probability of success
- Principles (Chapter 3)
- Values
- Program strategy and overall actions to meet business/mission objectives (from charter)
  - Safety and environmental
  - Operating/mission effectiveness
- Improvement process described in detail
- Organization, management, control, and administrative systems (full compliance with ISO55000 series)
- Opportunity analysis: specific opportunities for improvement to meet mission/business objectives, complies with program strategy
  - How determined
  - Calculated value gain by:
  - Contribution to business/mission effectiveness
    — Increase production: availability, production rate, and quality
    — Reduce cost
    — Improve capital effectiveness
    — Improve customer satisfaction
  - Effectiveness of physical operating assets
    — Improve operating efficiency
    — Improve reliability, cost, and maintenance effectiveness
    — Identify potentially disruptive conditions as early as possible: "eliminate surprises"
    — Maximize efficiency of all work processes
  - Organizational improvement
    — Improve safety and environmental performance
    — Elevate responsibility, build ownership, and improve attitude, morale, and commitment
    — Improve proficiency, qualification, and training: work quality, and effectiveness
    — Improve communications: intradepartmental, interdepartmental, and working level
  - Procedural improvements
    — Perform all work safely: without injury or significant incident
    — Increase work quality, improve repair success, and minimize rework
- Detailed improvement action plans (for each initiative)
  - Opportunity for improvement
  - Justification: gap to best performance and industry benchmarks

- Objective results and future state
- Strengths to build on and weaknesses and potential barriers to overcome
- Value delivered on completion: contribution to business/mission objectives
- Detailed actions and activities to achieve objectives
  - — Time and resources (technology, people, processes, and systems) required
- Measures of performance; metrics: activity and result, KPIs
- Probability of success and risk: actions to maximize probability, minimize barriers and risk
- Control, sustaining, and continuous improvement plan
- Assignment/responsibility for results

- Procedure for periodic program assessment and review (described fully in Chapter 17).

The program plan is a living document. It is ultimately created by Program Leadership and Improvement Action Teams, approved by the Steering Team, endorsed by executive management, implemented, publicized, available, and reviewed at regular intervals. An initial review should be conducted by the Program Leadership and Steering Teams approximately 6 months after program initiation and at any time program composition or direction appear to diverge from the plan. Regular, periodic reviews should occur every 12–18 months, thereafter with revisions made as necessary.

APPOINT WORKING-LEVEL LEADERS—CHAMPIONS

Every organization has hidden talent. There are people with the credibility, ideas, motivation, enthusiasm, and capability who are ideally suited to lead an improvement process at the working level.

> A key individual became seriously ill immediately before an important presentation. Lacking an obvious replacement with the requisite knowledge, managers had to consider all alternatives. An individual in another department had excellent technical knowledge but was very quiet, had never made a presentation of any kind, or been considered an advocate or leader. The replacement performed superbly—so well that a position was created to develop this previously hidden talent.

By its nature, an improvement initiative such as Operational Excellence is an ideal vehicle for identifying, encouraging, and gaining maximum contribution from people who may never have had the opportunity to use their full talents and capabilities. Identified by members of the Steering Team and Program Leader, many people with opportunity and encouragement suddenly blossom in terms of leadership and contribution.

From this group, the Steering Team and Program Leader will select a dozen or so as initial members of the Program Leadership Team introduced in Chapter 7 and to be described more fully in the next section.

Program Leadership Team

The Program Leadership Team (referred to as *Leadership Team* hereafter), along with derivative Improvement Action Teams, is the heart, soul, and muscle of an Operational Excellence program. Made up of working-level and technical supervisorial leaders, individual Leadership Teams should be formed for each operating unit within enterprises/sites composed of multiple, dissimilar operating units. Where operating units are similar or nearly so, for example, multiple units in a power generation facility, the decision whether to have one or more Leadership Teams should be based on specific circumstances such as similarity between units and applicability of improvement initiatives between units. As in all aspects of Operational Excellence, the decision to have one or more Leadership Teams is reviewed periodically and altered if changes will provide better results. In either case, the Program Leader is responsible for coordination.

Sought after attributes for potential members of the Leadership Team include the following:

- Credible within the organization: respected equally by peers, subordinates, and supervisors
- Competent, skilled, strong leadership capabilities
- Committed, belief in improvement, and the potential of the Operational Excellence program
- Positive, persuasive, enthusiastic, and energetic
- Open to new ideas and methods
- Works well with people, team builder, good coach, and gains the best from people
- Good problem solver and problem-solving skills
- Well organized and follows up
- Good communication skills and unafraid to express opinions up and down the organization
- Solid technical knowledge
- Trains successors.

Leadership Teams are appointed by the Program Leader in consultation with the Steering Team. Led by the Program Leader, leadership teams are the working groups responsible for most program operating details and key to success of the Operational Excellence initiative. As mentioned earlier, the Leadership Team is made up of technical professionals, working-level supervisors, and hourly people from all the functional organizations included in the Operational Excellence program. Members of the Leadership Team must be freed to devote at least 50% of their time to the Operational Excellence program during the important first formative months.

It is crucial to have the right people in the team. They must be highly respected with broad site/facility experience and knowledge. All must be committed to improvement, open to new ideas, and able to work productively in multifunction teams. Facilitation, at least early in the process, is advisable to gain maximum assurance that the mixed group will work together constructively and most effectively.

> At an introductory workshop, hourly people took seats on one side of the room—professionals and supervisors were on the other side. At another workshop in an international facility, participants divided along ethnic lines. During team discussions during a third workshop, engineers were leading all the working groups, doing most of the talking and recording notes. Hourly people were essentially silent.

In each of the case cited previously, action was necessary to break up the natural groups to assure maximum participation and expose all ideas while minimizing influence on the basis of perceived position in the organization. Action includes mixing supervisors, professionals, and hourly people who don't normally work together on leadership and action teams and, in extreme cases, forbidding professionals and supervisors from team leads or making group presentations. In this case, diversity is strength.

> In a working-level group, the individual chosen to make the group presentation confessed that in 20 years with the company, he had never before made a presentation of any kind. He made an outstanding presentation; enthusiasm, knowledge, and commitment came through clearly to all in attendance.

Activities for the Leadership Team typically begin with an alignment workshop similar to that described earlier for the Steering Team. The workshop is conducted by the Program Leader.

As part of the alignment workshop, the Leadership Team will review the initial program plan, mission, principles, and strategy developed by the Steering Team for conformance to objectives, practicality of improvement, and implementation. Specifically, the Leadership Team must determine whether the guidance provided by the Steering Team matches their knowledge and estimate of improvement opportunities to fulfill mission/business objectives. The team will identify and develop recommended improvements to the plan for refinement and later submission to the Steering Team. There may well be some back and forth accomplished by the Program Leader to gain full alignment between the Steering Team and Leadership Team's vision of what they believe is necessary and practical to meet business/mission objectives.

As specific initiatives for improvement are identified and defined, members of the Leadership Team will become leaders of multifunction Improvement Action Teams formed to pursue specific initiatives. Benefits are many. Most important, this pyramiding structure assures greatest use of organizational "brainpower," initiative, and institutional knowledge.

> In two separate facilities, engineers decided they wanted to become lubrication champions and establish a site-wide lubrication program that could contribute a great deal to operating effectiveness. Both personally organized comprehensive lubrication programs, established procedures, and conducted training that within about a year had generated substantial improvements and demonstrable value.

Once agreement has been reached on principles, the Leadership Team begins the actual process of identifying and value prioritizing opportunities for improvement as will be described in Chapter 12.

ADD/OPTIMIZE SUPPORTING PROCESSES

Now is the time to assure that all supporting processes necessary for full function and results of the Operational Excellence program are in place. Information and business systems are two primary examples.

Information Structure and System

A well-configured, secure, and accessible data, information, and reporting system is essential to identify and prioritize opportunities for improvement. It is also necessary to develop and convey the results gained by the Operational Excellence initiative in terms that will attract and maintain interest and support at the highest levels of an operating organization. If an adequate information structure and system are not in place at the beginning of program implementation, they must be created and implemented as soon as possible so that improvements establish a history of progress as they are implemented.

Data and information management systems must bring essential knowledge, data, information, and support to people charged with response. Information empowers creative people seeking ideas for improvement. Systems must be accurate, up-to-date, secure, safe from unauthorized alterations, accessible, and easy to use for people requiring information. Essential composition includes the following:

- Site/facility design drawings and specifications
  - Process (P&ID) and detailed drawings
  - Specified performance including operating limits
- Master equipment list (MEL)
- Operating and repair manuals
- SHE/EHS incidents and resolution
- Operating and maintenance history
  - Historical data and information
- RCA reports, including follow-up and resolution
- Program metrics and KPIs
- Document and drawing management system
  - Includes methods for security, updates, and changes to assure currency
- "Unofficial," alternate systems of information prohibited.

As a minimum, information systems must be able to:

- Separate and track improvements and costs by activity and asset
- Provide cross-correlation to enable performance, reliability, and cost comparison between similar activities and assets and even by original equipment manufacturer (OEM).

Information must be readily available to automatically calculate each optimization program metric and KPI and to create periodic reports communicating progress, performance, and results. The information system must be capable of collecting and consolidating data from disparate sources and supplying results for display in easily understood dashboard presentations similar to that illustrated in Figure 8.2.

Activity-Based Management/Accounting

Similar to the information system, a structure to accurately assess real costs and value produced must either be present at program initiation or be implemented soon thereafter to enable reporting real results. Today, most enterprises have this capability in their enterprise resource management (ERM) business systems. For the purposes of Operational Excellence, the structure should be reviewed to assure that results can be tracked down to the level necessary to identify opportunity and report value. ISO14224 specifies a nine-level hierarchy. A minimum seven seven-level hierarchy from facility to component as illustrated in Figure 9.3 is considered essential for a manufacturing/production facility.

A large facility organized its modern business system in a three-level hierarchy: plant, unit, and ID. Performance and costs of equipment assets couldn't be tracked below the unit ID level. As an example, the system could report costs charged to an equipment ID but was not capable of the essential differentiation between the drive or driven equipment, components such as bearings, seals, coupling, and auxiliary equipment.

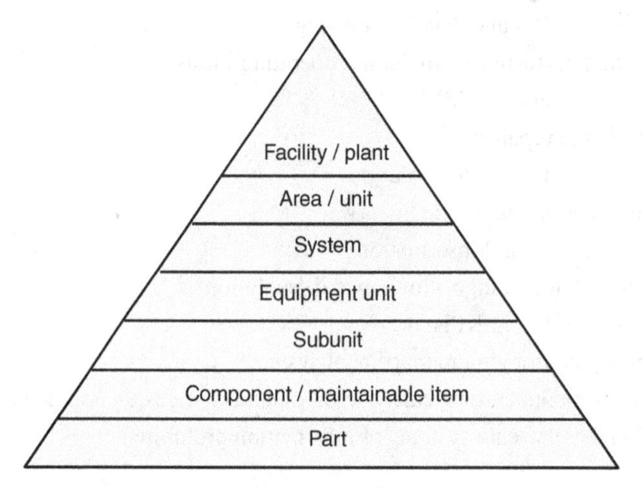

Figure 9.3 Typical hierarchical taxonomy, also refer to ISO14224.

Comparing operating and cost performance of similar components across the facility to determine lifetime value and spot poor performers was impossible. The shortcoming was highly limiting and eliminated the business system as a source of cost and reliability information that could illuminate and drive potential improvements. In this particular case, the IT department was uncooperative, stating that no departures from the standard three-level configuration would be permitted regardless of the limitations it posed in other areas. Work around efforts to gain vital performance information was similarly forbidden. In this particular case, the tail was literally wagging the dog!

ROLL OUT OPERATIONAL EXCELLENCE PROGRAM

With the program fully defined and mission, objectives, structure, principles, and leadership approved and established, the Operational Excellence program should be rolled out to the enterprise/site/facility as applicable. The roll out program begins by communicating the necessity, objectives, expectations, value, and benefits of the program in simple, brief messages.

The messaging, developed by the Communications and Leadership Teams, approved by the Steering Team, is designed to educate the entire workforce on the necessity for improvement, benefits of Operational Excellence, and convey a basic understanding of the program implementation and expectations for success. This will be discussed in greater detail in Chapter 10.

MAINTAIN MOTIVATION AND ENTHUSIASM

Executive and operating leadership must maintain focus, enthusiasm, and motivation on long-term enterprise/mission objectives. The tendency to shift into a short-term tactical focus where all efforts are concentrated on solving immediate problems must be resisted.

While maintaining the pressure necessary to gain short-term results, leadership must exert the discipline necessary to ensure that program tactics remain within the long-term strategy. Short-term fixes that are mandated by operational necessity must be followed by long-term solutions while knowledge is fresh. Point solutions must be leveraged with broader application wherever possible.

In a typical production operation, discipline is required to direct an appreciable level of attention to long-term improvement. There are generally so many problems requiring immediate attention that any reduction in intensity is welcomed and taken as a well-deserved break. Under these conditions, it takes solid leadership, continuously applied, to achieve long-term gains that may not be considered by those involved as contributing much to their short-term requirements.

Clear Acknowledgement That Sustainable Progress Requires Time

Leadership must recognize that improvement is not a command, but the result of a comprehensive, well-thought-out and planned implementing process. In numerous mid-level Operational Excellence workshops, participants state that management's belief that improvements can be ordered top-down is one of the most significant obstacles they have to face. Time and continuous effort are required to assure the process takes hold and is institutionalized.

WHAT YOU SHOULD TAKE AWAY

An Operational Excellence program must be specifically designed to meet enterprise and site/facility business/mission objectives. With that stated, the process and component parts are standardized beginning with the formation of a Steering Team and Program Leader. This is followed by creating a succinct mission statement, developing program-specific objectives and appointing the initial leadership to develop and refine improvement plans. Accomplished correctly, from here, success is essentially all about execution.

10

OPTIMIZE THE ORGANIZATION

At some point, early in the implementation of an Operational Excellence program, there must be a high level commitment to revise and optimize the organization as necessary. Reorganization to improve effectiveness, encourage ownership and responsibility, facilitate the formation, function, and success of multidiscipline improvement action teams, emphasize integrity, honesty, and trust, maximize human performance, and move business value-oriented decision making lower in the organization. Changes to an established structure are not easy and must be made with extreme care and sensitivity to assure that focus is on effectiveness within the principles and values of Operational Excellence, organizational improvement, and success, rather than change for change's sake.

All must recognize that most change, even when promoted as improvement, is unsettling. Change can disturb relationships, may infringe on what people think of as "mine" and what provides job satisfaction. Even worse, change may degrade conditions when change is based on assumptions rather than reality.

Operational Excellence represents a major change. Change in relationships and the way people are asked to think and perform, combined with greater demands placed on individuals. This type of change may be especially difficult within a highly structured, function-based organization. That's why strong, involved, engaged, honest and fair leadership, communications, listening, and follow-up are so exceptionally important to success.

Operational Excellence: Journey to Creating Sustainable Value, First Edition. John S. Mitchell.

ORGANIZATIONAL REQUIREMENTS FOR SUCCESSFUL OPERATIONAL EXCELLENCE

Assuring the success of an Operational Excellence program requires incorporating a number of identifiable organizational and human relations elements. Some have been mentioned; all will be identified and addressed in following chapters describing program implementation.

> A strong effort to achieve optimum organizational values and work culture at the beginning of a transformation initiative greatly facilitates gaining the necessary improvements in process and procedure. If the correct organizational values and work culture are established to build on, essential improvements in process, procedure, and habits will flow naturally. Attempting to improve processes and/or procedures, while leaving inconsistent organizational values and work culture to be addressed later in an improvement process, allows resistance to persist and significant barriers restricting success to remain in place—the initiative is likely to fail.

Inspiring People

Inspiring people to achieve their utmost and gaining consistent, maximum performance can be and likely is the greatest challenge to successful Operational Excellence. Teamwork and collaboration are musts. Operational Excellence must create an environment and working culture of shared vision, collaboration, ownership, and responsibility to gain the utmost performance. Many people have hidden talents. The search for organizational excellence begins by listening to people performing the tasks, what do they do, why do they do it, what are strengths, what are they proud of, and what are weaknesses where improvement can be made? Where multiple units/sites are performing the same activities, what are best practices, and where can best practices be combined to create "best of the best?" Listen, prepare a straw man combining best practices, get everyone in a room to discuss, and create a "best of the best" that all can agree on.

> An enterprise actively pursuing Operational Excellence identifies areas where improvement is considered necessary. They convene an internal "committee of experts" to identify and discuss applicable best practices. From that initial gathering, a "best of the best" practice is defined and sent to participants for comment. After several iterations with everyone more or less in agreement, the practice is disseminated. One participant who had worked at the company for over 30 years remarked that until participating in one such session he thought he knew everything possible about the subject. He was stunned to learn that there was much more to attaining best practice than he had imagined.
>
> A second enterprise follows a similar procedure. Preliminary best practices are posted on an intranet site for comment. After some fixed period that may include back and forth discussions between the "owner" of a best practice and people who will be responsible for implementation, a final version is posted for approval. Once approved, the best practice is incorporated into the operations procedures.

Consistency of Message and Action

Within an Operational Excellence program, consistency between message and action is imperative. There is either a commitment to safety, excellence, integrity, quality, and value or there isn't. Any compromise for expediency or any other reason will diminish commitment throughout the organization. There can't be any compromises with safety or integrity; however, in the real world, there may be other areas where occasional necessities for compromise occur: between value and excellence as one example. In this situation, it is imperative to describe exactly the conditions and why compromise is necessary.

> During a morning production meeting, the necessity for quality was discussed at length and in detail. After the meeting, a unit production superintendent accompanied the unit maintenance superintendent to the maintenance shop where equipment was being repaired. After the maintenance superintendent had described the repairs in progress and estimated time necessary for completion, the production superintendent began pressing for shortcuts that could be taken to expedite completion, even though the equipment wasn't required so long as the spare operated satisfactorily. Assuming the very best, the term quality and its explanation had two completely different meanings for two individuals hearing the same words at the same time.

Necessity for Improvement, Requirements, and Overall Objectives

As has been stated, executive management must establish overall requirements, results that must be achieved, a vivid vision, and clear concise objectives that can be understood by everyone. Requirements must be as specific as possible: produce more and spend less, preferably expressed in terms of safety, environmental, production output, cost, etc. with measurable percentage improvement objectives. Messaging must be positive: why are we doing this, why is it necessary, and what are benefits to you? The latter, benefits must be stressed. This topic will be addressed in more detail later.

> To achieve success, people must be provided a persuasive reason for improvement. There must be a compelling, believable reason for improvement, participants must feel ownership and responsibility, and there must be tangible rewards for success. All must be addressed to achieve success.

In addition to establishing necessity and requirements, the vision and objectives must direct attention to factors that contribute directly to enterprise, facility/site, organizational, and operational objectives; convey excitement; and create ownership for success. An easily accessible Frequently Asked Questions (FAQs) communiqué format, along with continuing communications emphasizing a point or two, is highly useful to keep essential messages in front of people. The communications must be kept fresh. FAQs are added as questions arise to assure that everyone is informed, positive, and engaged. The objective is not simply to state requirements but to get everyone personally and passionately committed to success.

A project was initiated within a multiple site enterprise to standardize procedures. Realizing that each site had its own way of doing things for which they felt worked well and were justifiably proud, it was recognized that success was dependent on selling the benefits of standardization to a reluctant audience. Benefits communicated in a FAQ format included a collegial process to meld individual best into enterprise-wide best practices, gaining maximum safety and quality from agreed on best practices and an improved ability to share successes and lessons learned. Offering promotions and transfers between sites without a lengthy learning period and the certainty, stability, efficiency, and happiness produced by standard processes are added benefits.

Benefits Operational Excellence benefits the enterprise in terms of improved SHE performance, business/mission effectiveness, and value. At the working level, benefits gained from Operational Excellence include increased:

- safety, reliability, cleanliness, and stability in the working environment
- pride, trust, ownership, contribution, personal satisfaction, and quality of the working life
- job stability and security
- professionalism across the entire enterprise: can do improvement-oriented work culture
- empower and influence others
- voice in processes and operations, greater control and consistency, and fewer surprises
- opportunity, to gain additional knowledge, achieve full potential, and learn and apply new capabilities and skills
- compensation based on contribution and results

All must be identified, publicized, and continually reinforced to maintain enthusiasm, ownership, contribution, and support.

Empowerment

Responsibility, empowerment of the individual, and accountability for results are important elements of Operational Excellence that are mentioned in most improvement texts.

Empowering the work force and demanding more and greater results and effectiveness are a competitive necessity in today's operating climate.

Although it is typically true that most people want to be listened to and considered, not everyone wants the ownership, responsibility, and accountability that accompany empowerment. And that's fine; not everyone can or should be a chief! Essential elements include that employees have the opportunity to achieve within their own comfort zone and work to complement each other. Teams must work well internally and with other teams. Within a team, there is plenty of opportunity for individuals to contribute, even though they may not want to be a leader or empowered. Leadership, coaching, and counseling must be available to identify and achieve individual objectives.

Establish the Basis for Organizational Improvement

Recognizing that the necessity for organizational changes will likely emerge later in the process, initial executive communiqués must prepare the ground by stating commitment to improvements throughout the enterprise, including organization.

> During implementation of an improvement initiative, it became apparent to most involved that a reorganization including altering the division of responsibilities was a necessity. Despite recognition at the working levels, middle managers appeared unwilling to consider any changes that they concluded might diminish their position or authority. Lacking executive commitment, possibly even awareness, the improvement initiative could not move as fast or as far as it should have. Initiatives that should have been combined to gain maximum results often proceeded in parallel with interactions and communications to coordinate and strengthen efforts forbidden.

Chances are that the need for reorganization will initially emerge from the working levels as inefficient processes, and procedures are identified for improvement. Some examples have been identified. In many cases, achieving the level of cross-functional cooperation essential for Operational Excellence will be resisted by mid-level managers who feel their control threatened as in the last example. A method must be established to identify and surmount this potential barrier to improvement. The Steering Team, introduced in the previous chapter, will be a key element and must initiate and champion reorganization where appropriate.

> A highly experienced engineer was accustomed to providing callout support to Operations on a 24/7 basis. Eventually, his manager forbade response without his personal authorization. In many cases, the manager wasn't available or wouldn't provide authorization, placing the engineer in an impossible situation. Ultimately, the conflict became irreconcilable and forced the engineer to leave the company.

Organizational Types Broadly speaking, there are two types of organizations. Command and control where everything emanates from above and subordinates do exactly as instructed is one type. There is very little consultative feedback. Essentially, "the boss is always right."

The command and control organizational "silo" structure in which workers are expected to stop thinking and perform exactly as they are told—no more, and no less—is a relic of the past. Silo thinking that shifts ownership and accountability to the supervisor and prohibits work across functional boundaries cannot continue. Today's supervisors are responsible for too many people and have too much administrative work to be able to effectively supervise every activity.

> Some years ago, an employee at a major U.S. automobile manufacturer commented that relations between union workers and management were highly adversarial. Rules

required workers to adhere strictly to the contract in terms of both work scope and performance. He stated, "management seemed to think that workers checked their brains at the time clock." Suggestions for improvement were not encouraged and certainly not considered. Eventually, the company was forced into bankruptcy. Although not the sole reason, the attitude represented was certainly a contributor.

The second is a consensus organization where decisions rely on mutual agreement. It resolves most of the problems cited previously. Although people may feel better working in a consensus organization, it is often difficult and time-consuming to gain consensus. Furthermore, the consensus position may represent a compromise that is not optimal.

A hybrid-type organization may be the best and certainly should be considered for Operational Excellence. In a hybrid organization, individuals are delegated responsibility for decisions. For the organization to function most effectively, the decision maker must listen to and consider all viewpoints and recommendations. Once a decision is made, subordinates must agree to fully support the decision, with best effort applied toward success, regardless of whether they favored the course of action. Efforts to make the decision fail, sabotage, are not tolerated. As action proceeds, the decision maker is obligated to continually review the decision and make corrections where necessary, even if the changes prove correct those who were in favor of another course of action.

The hybrid-type organization combines the best of all worlds. Decisions can be made quickly. Everyone is obliged to support the decision maker who, in turn, is obliged to act responsibly and make improvements where necessary, even when the improvement goes against the original decision. The hybrid type of organization best fits the optimum work culture values detailed earlier.

To function most effectively, a hybrid organization requires fair, enlightened leadership throughout. Leaders must listen to and consider all recommendations and correct their decisions even if they fear that doing so makes them look weak. In fact, the opposite is true. A willingness to correct mistakes makes everyone stronger. There must be a means to identify, counsel, and ultimately correct leaders who abuse their decision making authority by evolving into a command and control, authoritarian style.

Organizational Principles Within Operational Excellence, organizational principles, considerations, and elements are directed to achieving the utmost value from human capital. In the modern world, the objective is gained largely by harnessing brain power through competent, enthusiastic employees who know their job and are continually thinking about and recommending how it can be improved. Guiding principles for the organization include the following:

- Streamline and flatten the organization for greatest effectiveness and responsiveness.
- Assure full connectivity and understanding between program leadership and those charged with implementation. The top-level leadership, Steering Team,

must understand challenges and problems at the working levels and display an equivalent commitment to the program and success as they expect from subordinates. A full commitment to success at all levels is imperative for program success and sustainable results.

- Increase the energy, ownership, teamwork, integrity, trust, responsibility, accountability, and effectiveness of all personnel.
- Delegate greater responsibility and control to the lower levels of the organization and shift behaviors to ensure greater individual initiative, ownership, and accountability.
- Energize positive contribution and improve relationships, communications, teamwork, and response to deficiencies.
- Reduce non-value-added activities, unnecessary bureaucracy, and layers of management required for approval.

Guiding principles must address improvements required within an evolving new organization. What strengths should be retained and built upon; what weaknesses need correction? What improvements are necessary?

Core values: safety, integrity, honesty, commitment, trust, discipline, and good communications are characteristics of a top-tier organization and essential building blocks for improvement.

All organizations have a different institutional culture, values, relationships, individual personalities, behavior, and controls. The listed principles are universally applicable; however, there is no single, one size fits all, solution to achieve excellence. Using overall principles, values, and individual initiative and allowing flexible implementation will build ownership, enthusiasm, and initiative.

INITIATING ORGANIZATIONAL IMPROVEMENT

An ambitious organizational improvement program may involve a great deal of change from the current work culture, practice, and organizational structure. As a result, many may feel uncomfortable and resist the Operational Excellence program.

Improvement Management Process

Depending on the internal working atmosphere, enthusiasm, and commitment for improvement and the Operational Excellence program, it may be necessary to augment the process by creating an improvement management process accompanied by formally appointing a change manager or improvement leader. Since the task is associated with duties of the Communications Team to be described later, the improvement leader should be the leader or at least a member of the Communications Team.

The improvement leader does not have to be a full-time position. However, the individual must have sufficient time to interact with people at all levels of the organization whose enthusiastic support is essential for success. The position requires face-to-face discussions; e-mail, text, and other impersonal communications are not sufficient.

The appointment of an improvement leader should shift the activities of the Communication Team to proactive in nature where the improvement leader and team members will actively solicit comments, questions, and concerns regarding the Operational Excellence principles, values, requirements, and program and make every effort to ensure questions, concerns, etc. are resolved as quickly as possible.

Identify Potential Structural and Organizational Improvements

A reorganization plan may develop as the Operational Excellence program evolves, and opportunities for organizational improvement emerge in the process. While outside facilitation may be very useful to speed and optimize the process, inside people must be involved who are intimately familiar with the current organization, culture, conditions, values, strengths, individual personalities, informal relationships that make things function, weaknesses, potential barriers, and opportunities for improvement.

Organizational improvements may be addressed in workshops to be discussed in the next chapter. If so, the workshop must include a broad cross-section of people within the existing organization (professional, managerial, and supervisory: salaried and hourly) so that all problems, structural defects, potential barriers, and opportunities for improvement will be identified and examined from every perspective.

Everything should be "on the table" for consideration. This includes roles and responsibilities of essential functions and their location and reporting within the organizational hierarchy.

Move the Organization from Management Control to Team Partnership

Within Operational Excellence, the organizational improvement process typically evolves from multifunction work teams. As teams are empowered to identify and correct deficiencies and demonstrate success, a hierarchical command and control organization is gradually augmented by a flattened multidiscipline team organization. This is a logical step in that people directly performing work are most aware of deficiencies and problems. Given the chance and properly supported, they will come up with innovative improvements to overcome deficiencies. Teams gather all with common knowledge and have responsibility for defining improvements, ownership and responsibility for implementation, and accountability for results. Direct supervisors become less operating foremen evolving into a role of facilitation, removing barriers, coaching, and monitoring results.

Building individual and team capabilities within a functional organization is a prime role of leadership. With time, teams gain greater self-reliance and direction, control over tasks, prioritization, and accountability for results. Leadership assures teams' progress by establishing and monitoring performance measures and effectiveness KPIs, training and coaching, respecting and considering team recommendations, and explaining deviations when recommendations can't be implemented for any reason. The improved organization will probably be a matrix, with people having both functional and team responsibilities. Functional and team leadership must work out details of time allocation to optimize results in both areas.

It should be noted that it may take several years to get a team concept fully aligned and working effectively.

Work Culture

The work culture within an operating enterprise is top-down. Work culture originates with the honesty, integrity, consistency, and example set by leadership and management. It governs the actions of all employees. An optimum work culture respects people, demands honesty and integrity in all activities, and encourages positive behaviors and pride. Solution of problems and issues occurs through a collaborative team effort at the level on which the problem exists with all relevant facts and information available. Knowledge gaps are bridged through cross-functional teaming.

All within the organization must abide by the honesty and integrity guidelines. This extends to equality, not showing favoritism for any element except competency and commitment. Supervisors who are more generous in areas such as overtime, assigning easier tasks, and even days off to "friends" seriously erode a work culture that demands honesty and integrity.

Achieving Individual and Organizational Success

Identifying and encouraging competent, motivated, and fast-learning individuals, giving them as much responsibility as they can accept competently, and moving them into influential positions as rapidly as possible are key elements toward achieving an improved organization. Some individuals will thrive. These individuals must be given training and increased responsibility as quickly as possible. Building on success in this manner self-perpetuates a new, improved organization.

> "Positive attitudes and behaviors in industrial operations are the key essentials to gaining success"
>
> Dr. Peter G. Martin, Vice President; Invensys

When individual success is combined with team-based rewards, peer pressure becomes a powerful incentive to elevate ownership and performance. Recalcitrant members must be urged to participate fully and raise their performance to team average as a minimum. That said, they may be potentially strong contributors who simply do not fit the new team model. In this case, effort must be made to find ways to preserve and reward their contribution.

> Workshop participants engaged in improvement initiatives frequently state that senior management often identifies the necessity for new behaviors and demands performance

to increase value and results. At the same time, performance is measured and evaluated by old activity-based metrics. Participants laughingly question why there is any question that behavior doesn't change.

Typically, the best professional and working-level people are promoted into supervisory roles as a reward for performance and contribution. Some, even with extensive training, prove deficient as supervisors. Rewarding individuals, both monetarily and with self-esteem, who may be excellent contributors at the professional or working level but deficient as supervisors, is a challenge that must be overcome.

Improving Behaviors

There will always be laggards, people who resist change and improvement, regardless of how positively it is presented. For people in this group who are otherwise solid, productive employees, counseling, and additional training may be indicated. Grumbling and complaining are characteristics of any operating unit, no matter how good. Some of this can be laughed off, as everyone knows who the grumblers and complainers are. However, negativity must not be allowed to permeate the working culture. Behaviors and comments negatively affecting morale must be stopped in their tracks. People who state publicly they are in favor of an improvement initiative while working behind the scenes to create resistance and sabotage efforts are absolutely intolerable and cannot be allowed to persist. It must be made clear to all that once a decision is made, everyone must be fully supportive, as that's the only way to identify shortcomings and flaws. Likewise, everyone involved must be assured that if unanticipated flaws and shortcomings are discovered, timely corrective action will be taken.

A key, highly valuable individual involved with an improvement initiative had a large personal investment in one element of the initiative. When that element was rejected, he became negative and sullen regarding the entire initiative and the people responsible for the rejection. It was made very clear that while he was fully entitled to his opinions, attempting to sabotage the overall initiative with negativity communicated to others would not be tolerated. To his great credit, he did not transmit his personal feelings or opinions into the organization.

SKILLS MANAGEMENT AND TRAINING

Creating the work culture and skills necessary for successful Operational Excellence requires diligence, management, and targeted training. Skills management assures that skills, proficiency, qualifications, and certifications required to implement and sustain all activities within Operational Excellence and associated programs are totally defined, documented, and up-to-date. Furthermore, adequate personnel must be available at the required levels of proficiency to meet requirements. Skills management includes time and funding for a comprehensive training program to assure proficiency. In addition to skills training, the program will include SHE, team, organizational improvement, and related areas. The training program must include auditable records of training completed, certifications, and retraining requirements. A skills matrix identifying personnel qualified to specific requirements as well as alternates for back up in the event of absence is essential.

Operational Excellence training is conducted in a highly focused manner into specific areas requiring knowledge. Training ranges from a summary overview to acquaint executives and managers with program function and objectives to very specific, detailed training for people who will be directly involved with the Operational Excellence program implementation and/or a specific procedure. After completion of training, continuing opportunities to exercise skills to assure and maintain proficiency must be provided.

Skills and other training must be conducted from processes and procedures in actual use amplified by training manuals as necessary. It is absolutely essential that training, processes, systems, and procedures are totally aligned and synchronized with how work is actually performed. The common refrain communicated to a recently trained individual reaching the working level "forget everything you were told during training; now let us show you how this task is really accomplished" must be avoided at all costs.

Information as a Basis for Training

Properly established, the enterprise information system is a vital source for identifying the necessity and objectives for process, technical, and practice training. While care must be taken to prevent misuse, information that indicates the success and effectiveness of technologies and practices is quite useful. Repair success, a metric that indicates a given repair is accomplished on schedule and successfully restored to service, is one example. A low repair percentage success for a given task can indicate a design problem, inadequate procedure, necessity for additional training, or combination of the three. In any case, it reveals the necessity for further study. Equivalents can be used for other processes.

This idea can be taken a step further to the evaluation of personnel competency. Here again, it is essential to stress that information developed for this purpose must be used in a constructive and not punitive manner. Constructive to indicate a need for skills training, perhaps even assignment of work tasks. Proficiency information must not, under any circumstances, be used for disciplinary purposes with a corresponding likelihood that potentially valuable information will be skewed.

> An industry-best facility used a work task success metric to direct the necessity for additional training and even work assignments. The latter to assure competencies and demonstrated performance were best matched to assigned tasks. As a unionized facility, they fully recognized the potential discord such a system could initiate and therefore took extreme caution to assure the system was used positively.

Refresher training must be implemented in at least three areas: the program itself, infrequently accomplished processes and activities, and areas where continuing training and recertification are required by regulation and/or procedure. *Training* is also called for in areas where performance and results indicate the necessity for additional training. The activity itself and recognition of need through assessments will determine the scope and interval between training.

Building a "bench" of qualified people is another area that generates training requirements. Oftentimes, a site or facility will have only one person fully qualified for a specific task or activity. If that person goes on vacation or becomes unavailable for any reason, the task can't be performed with the necessary requisite proficiency. When the task requires certification, the necessity for an alternate is even more important. Thus, it is essential to have multiple people trained and fully qualified to perform all tasks and activities to avoid total dependence on a single individual. In this regard, training alone is not sufficient. To retain full qualifications and assure proficiency, potential substitutes must actually perform the task or activity at regular intervals.

An electronics manufacturing company allowed total competency for a critical production process to reside in a single individual. When that individual contracted a long-term illness, production of a vital component essentially stopped with major impact on revenue and customer satisfaction.

PERSONNEL REDUCTIONS

Like it or not, many, if not most, people within an organization will interpret any initiative to increase efficiency/effectiveness as a means to get rid of employees and/or reduce compensation. Resolving this inevitable conclusion as early as possible during the initiation of an improvement program is probably the single greatest factor to build enthusiasm, ownership, acceptance, and support.

The issue must be met head on with solid, credible answers before concern has a chance to grow into a major barrier to acceptance. As one of the first actions at the commencement of an improvement program, the best enterprises map out an organizational structure and expected employee count at objective levels of efficiency and operating cost. The projection goes down to the work group level. The future number of employees is always less than the current.

The next step is to look at demographics and historical attrition data to identify the number of retirements, transfers, resignations, and separations likely to occur during the program planning period, generally 3–5 years. If the organization is fortunate, and many are, simply not replacing retirements and normal attrition will produce the projected employee count when the program initiatives are met. No layoffs will be required. If they are not so fortunate, some early retirements or layoffs may be necessary. In either case, it is far better to know and communicate the answer with a credible action plan than attempt to dodge a question that can absolutely destroy an otherwise solid improvement initiative.

One facility conducted a demographic study at the initiation of an improvement initiative. They found retirements and normal attrition would result in an employee level that was actually below objective. Since the study included a detailed organizational structure by position, they could quickly ascertain where shortfalls would occur. As a result of this study, they were able to announce to all employees the very good news that head-count projections would be met by normal attrition; there would be no layoffs. It also led to the initiation of an apprentice program in craft areas where shortages were projected to occur.

A second facility embarking on an improvement initiative totally defined their objective organization down to the work group level. They accomplished a demographic study to identify potential reductions by position and work group. In some cases, they initiated cross training so that loyal, solid employees who might be vulnerable to downsizing because of skills and membership in a specific work group would have more flexibility and thus be retained in a slightly different skill.

A third site found the number of highly skilled, certified crafts necessary during outages well above what was needed for normal operations. Plant management gave people in these categories a choice: learn additional skills that were necessary during normal operation, the facility would provide the training, or accept temporary employment. Most chose the former.

At another site, an estimate of attrition, conducted at commencement of an improvement initiative, disclosed the site would be slightly over objective employee count at the end of the first phase of the initiative. The site decided not to take any action beyond not replacing most retirements, transfers, and resignations. The decision was announced to the great relief of all employees. Midway through the program with objectives partially realized, the enterprise ran into cost problems due to a general economic downturn and competitive production emerging from a lower cost area of the world. Major layoffs were directed to all sites. Since the site with the improvement initiative had not replaced retirements and resignations for over 2 years, these were credited against the ordered reduction in force with the result that additional layoffs were minimal.

NECESSITY FOR EFFECTIVE COMMUNICATIONS

Communications are one of the most important aspects of an improvement initiative. They are perhaps the greatest need and toughest challenge and also frequently one of the most ignored. Communications form the vital link between employees, leadership, requirements of the enterprise, site/facility, and the Operational Excellence program.

Communications providing employees with knowledge of the business/mission environment and the necessities for improvement to assure continuing success and job preservation are crucial to promote participation, ownership, and a commitment to excellence within the Operational Excellence process. If most aren't convinced of the need for improvement—no improvement will occur!

Communications are an information, promotional, and feedback tool. Communications are absolutely vital to establish and maintain the internal energy, enthusiasm, ownership, commitment, and constructive conditions essential for the improvement initiative to succeed. They create awareness and understanding of the necessity and benefits of Operational Excellence and keep the objectives and benefits of improvement uppermost in everyone's mind.

Some assert that it is better to keep people largely in the dark when faced with the necessity for process and organizational changes that many might find highly threatening. Others argue that communicating frankly and addressing concerns as they arise is far better than attempting ad hoc answers to rumors that fly whenever any uncertainty or unexplained activities are in the air. The second group is usually the most successful!

A site manager responsible for a large improvement program felt that too much knowledge and communication would upset employees. He therefore restricted his objectives, plans, and proposed actions for improvement to a small circle of senior managers.

Effective communications go far beyond published and visual material. Communications must be an interactive process to create the values, attitude, and environment that are essential for consensus and success. For maximum effectiveness, communications must be simple, consistent, and repetitive. Getting a vital message embedded into the work culture requires repetition: each iteration from a slightly different and complementary direction.

During one workshop, a consensus emerged that complicated communications such as changes to employment and/or benefit policies should be communicated to the working level by people intimately familiar with the policy rather than group leaders who may not fully understand or be able to explain all the issues and details involved.

A formal communications program is a key element for building the trust necessary for success. It must be directed to explain the necessity and benefits of the Operational Excellence program, convey straight answers to reservations, concerns, and questions in simple terms, and publicize and promote success. Specific examples of successes and lessons learned are of special interest and convey highly credible information that is immediately relevant.

Communications Requirements

Communications requirements must be formulated and built into the Operational Excellence program from the outset. Communications requirements must specify who to communicate, what to communicate, and how messages are to be communicated. A rigorous method of follow-up must be instituted to assure questions, reservations, and concerns are quickly identified and resolved.

A comment during a workshop stated that while communications at the upper strategic level appeared OK, and alignment looked good on paper, a recent survey indicated that only about 50% of supervisors were aware of and understood the objectives, KPIs, and reasons for the improvement initiative.

What to communicate, messaging, is essentially determined by the specific group needs. Progress and success, high level financial and operational results and benefits,

Category within program:		Communicate to:	Receive from:
Responsible *Steering Team, Program Leader*	*Management responsibility for program, direction, follow-up and achieving successful results.*	*Program and Initiative leaders: Requirements, program objectives, benefits. All (through communications team): Program necessity, requirements, benefits. Results and successes.*	*Program and Initiative Leaders: Results, successes and contribution to enterprise objectives in increased business value / mission effectiveness. Barriers, challenges to success.*
Accountable *Program and initiative leaders*	*Accountable for opportunity identification, plan, implementation, follow-up and success*	*Improvement action team members: Objectives, program essentials and requirements for improvement.*	*Steering Team: Objectives and requirements for improvement.*
Support *Senior executives, facility/site and business managers*	*Provide vision, strategic guidance, direction and continuous support. Remove barriers to success.*	*Steering Team: Vision, business / mission objectives. Anticipated changes in business / mission requirements.*	*Steering Team Results, successes and contribution to enterprise objectives in increased business value / mission effectiveness.*
Consult *Typically support personnel; supervisors, Subject Matter Experts, HR, IT, etc. included for specific expertise and/or in a position to affect progress*	*Peripheral involvement with execution that may involve additions to work routine.*	*Program Leader, Steering Team Friction, potential problems.*	*Communications Team: Overall program requirements, objectives, successes and benefits. Program Leader: Activities that may require participation and support.*
Inform *All personnel*	*Observe program activities; not directly involved in implementation.*	*Program Leader, Steering Team Friction, potential problems.*	*Communications Team: Program necessity, objectives, activities, successes and benefits.*

Figure 10.1 Example communications matrix.

and contribution to enterprise business/mission objectives are communicated to executives. Necessity, benefits, progress, and results are communicated in brief form to all in the enterprise to generate interest and support. Implementing details, results and lessons learned, are communicated to those accountable and responsible for program results. Figure 10.1 illustrates a typical communications matrix for the five groups identified in the following section.

Communication Groups

There are at least five groups for which communications are developed. High level executives and managers whose participation, support, and approval are essential to success are in one group. People and groups accountable for results are in a second group. All directly involved and responsible for identifying and implementing improvements are in a third group. People who might not be directly involved but are in a position to affect improvement activities are in a fourth group. Finally, there are those who may observe program-driven changes and ask why they are being made. Messages to all groups must be consistent and in agreement; however, emphasis will be quite different. For convenience, the grouping is delineated in a loose RASCI form introduced in Chapter 9.

Responsible: all engaged in the implementation, follow-up, and achieving successful results from the improvement effort

Accountable: primarily program and initiative leaders

Support: senior executives and managers including the Steering Team: primarily results, successes, and demonstrated contribution to enterprise objectives in terms of increased business value/mission effectiveness

Consult: influencers, not directly involved but who might be in a position to affect progress

Inform: observers, not directly involved but must be aware of program objectives, activities, successes, and benefits.

Note that communications to different groups may have many similarities. For example, communications to the Consult and Inform groups may have far more similarities than differences. Furthermore, some groups, the Steering Team for example, may receive selected communications directed to multiple groups depending on the specific details. Those directly engaged in the program should receive communications directed to all groups for information and to demonstrate the importance of the Operational Excellence program and results.

Responsible While communications from the following groups are primarily inward toward the receiver, communications with those actually engaged in identifying, developing, and implementing Operational Excellence improvement initiatives will be two way. Program implementer communications will be primarily with program leadership, including the Communications and Steering Teams. Information will include program and specific objectives, description and status of working, and proposed improvements, benefits, and value generated for each. Progress to milestones and any challenges to successful completion for improvements in the implementing stages will be included. Successes and lessons learned are always emphasized.

In this group, feedback is all important and must be continuously solicited. Uncertainties and questions all must be sought, addressed, and communicated as soon as possible.

Accountable This group is primarily program and improvement initiative leadership. (For purists who demand sole, individual accountability within a RASCI, the Steering Team, who are probably considered accountable at the executive level, can be moved into the previous, Responsible category.) Communications begin at the inception of the improvement identification process where leadership must be briefed on the scope of potential opportunities, their value contribution to enterprise, and program objectives, resources required, and probability of success. The Steering Team will review and approve plans and monitor progress to objectives. A formal structure must be established to present essential information in a concise format meeting all requirements for monitoring progress and identifying departures in sufficient time for corrective action.

Support: Executive Management Regular communication is imperative to create and sustain essential support and drive for the overall Operational Excellence program and improvement process. Communications are at a high, strategic level, primarily conveying plans, status, results, and benefits emphasizing increased business value/mission compliance to include the following:

- Improvements in safety and environmental performance
- Contribution to enterprise business/mission objectives: production/mission compliance and reduction in costs
- Improved operating performance and effectiveness, compliance to operating/mission requirements, reduction in operating interruptions, unexpected failures, and other events of direct interest to executives
- Successes, especially where improved safety, environmental performance, and operating effectiveness are a direct result of the Operational Excellence program

Consult: Influencers; Inform: Observers Communications requirements for people not directly involved with Operational Excellence are, for the most part, fulfilled by a brief "elevator speech" type of communication describing necessity, benefits, and results. As the Operational Excellence program matures, communications will shift to concentrate on results and successes supplemented with briefs addressing subjects of specific interest.

In addition to why and benefits, some influencers such as site/facility staff will require more detail to assure support. Specifics are dependent on and should be targeted to position and duties. As an example, it may be necessary to justify the necessity and benefit for recording detailed downtime and failure data to an Operations or Maintenance supervisor. Benefits for the individual must begin with What's In It For Me (WIIFM). Without a convincing answer to this question, there is very little chance the requirement will be fulfilled accurately and consistently.

> Many have commented that requirements expressed for additional information must be accompanied with communications explaining why necessary, how it will be used, and its benefits. Any discoveries made with the information supplied must be shared with those who have been the suppliers as soon as practicable along with value and benefits gained. Feedback of value and benefits is absolutely necessary to sustain the motivation for collecting accurate data and information.

In the staff area, people will ask for detail that pertains to a specific area of interest such as how does Operational Excellence apply to and affect engineering, HR, finance, IT, and possibly back office processes. All must all be aligned and kept informed by means of periodic communication and reports to ensure they are fully supportive of the Operational Excellence program: prepared to participate and contribute when needed.

During a discussion of potential improvements requiring additional training and advanced skills, it became apparent that union rules and task certification would have to be considered in order to arrive at an acceptable solution. HR involvement was imperative.

Reviewing some inconsistencies in another improvement initiative, it was recognized that a large part of the solution resided within Business Services. They were flattered to be asked to participate and offered full cooperation.

Messaging

The basic message must be constructed to assure that all employees understand the necessity for improvement and visualize how they must contribute to and will benefit from success. The focus on changing attitudes and beliefs is communicated by stressing necessity, objectives, results, and benefits. The necessity for improvement (why we must improve) must be thoroughly explained along with the destination and journey (where we are going and how we get there).

Specifics to be considered for inclusion in the messaging include the following:

- Necessity for improvement to meet changing business/mission/market conditions
- Benefits for employees including improved job security and satisfaction
- Summarized major principles including safety, honesty, trust, excellence, and value in all activities, broad participation, commitment equivalent to safety, and continuous sustainable improvement
- Organizational values beginning with fair, open, honest interactions with all employees
- Clear description of the inefficiency, cost and personal impact of unnecessary/low value work, and preventable disruptions to operations
- Program implementation: what you can expect
- Transition plan: how we get from where we are to where we need to go
- Potential concerns including reduction in employment and changes in responsibilities
- Encourage questions and comments.

Every effort must be made to assure that all communications are consistent and complementary with the sole difference being the depth of knowledge required to assure enthusiastic acceptance and support that will produce success and sustainability of the Operational Excellence program

Trust is built from action and deeds; communications are a method to disseminate information and results that build trust. Actions that are visibly consistent with values, enterprise, and program objectives, honest and fair to all, are prime opportunities for communications. Communicated effectively, this will build trust and confidence in the process and in the people responsible.

Communications Program

A communications program is a key element for building the trust necessary for success. It must be directed to conveying information and straight answers in clear and simple terms that will be understood. Limiting communications to a single subject and the use of metaphors that connect to everyday experience are good methods to assure that complex ideas are understood.

All communications must be checked before release to ensure that they address a valid subject, are accurate and clear, and will be fully understood by everyone who needs to absorb the information. In all cases, questions must be encouraged and a methodology provided for timely, thorough answers that are communicated broadly as applicable.

Communications Team

The formation of a communications team early in the implementation of the Operational Excellence initiative is imperative. The communications team is appointed by the Steering Team and tasked to assure that the communications already mentioned, as well those that will be found necessary throughout the improvement process, are developed, coordinated, positioned, and transmitted most effectively. Other vital functions include identifying issues, questions, and concerns that need to be addressed and answered to maintain a positive, participatory environment. The communications team will assure prompt answers to head off concerns and reservations that if left unanswered could diminish commitment and support. Communications and the communications team will gain importance as the initiative moves along.

The communications team should represent a cross-section of facility personnel. Members should include management, the professional staff, production, and maintenance personnel: salaried and hourly. Members should be selected for their sense and knowledge of the facility, people and culture, credibility and confidence within their peer group, willingness to contribute, and eagerness to improve conditions. They must be people with a good feel for the mood and spirit of the facility and in whom others confide. They must be willing to listen to a lot of complaining with the ability to sort out real from imagined concerns and be able to find answers for both. The communications team can be considered equivalent to a plant ombudsman.

> In every facility and every organization, there is a person or persons in whom others confide and has unique knowledge about the "pulse" of the facility and real and imagined problems. It is imperative that these individuals are identified and become members of the communications team.

The communications team works alongside the Steering and Leadership Teams to develop the communications necessary to promote the process and its benefits to employees, many of whom may be skeptical and concerned. The communication team's initial task is to develop a program communication strategy and plan that will assure that the entire organization is aware of the Operational Excellence program: its necessity, objectives, and benefits. The Steering Team must be involved in this process, as many of the initial questions regarding "why is this necessary" will require business/mission-based answers that must be supplied by the Steering Team. The communications plan should be directed to developing a consensus in favor of the Operational Excellence program. It must broaden support and dispel doubts about hidden objectives of which the first will be personnel reductions mentioned earlier.

Ongoing tasks include identifying and framing real questions and presenting the questions to the appropriate authority for answers. The communications team doesn't have to answer questions; their primary task is to identify, frame, and pass on real questions. The communications team makes certain the question and its importance are understood, answered completely by the appropriate people within management and/or Steering Team, and communicated back within a reasonable period.

It is crucial to emphasize that the primary function of the communications team is to assure timely communications that convey necessity and benefits, especially the latter, publicize successes, and answer questions. To restate an important function, the communications team is not responsible for providing answers but rather assuring questions go to the right people or groups, answers are forthcoming, and are publicized to all interested parties.

Methods of Communication

Methods of communication are many and varied; selection depends on the message, people who need the message, and facility practice. A company intranet provides broad and rapid delivery, but is it regularly viewed and, much more important, are messages taken to heart and acted upon?

> People will occasionally suggest placing an obvious spelling error in a message broadly communicated through a means such as an enterprise-wide video display to see how many are paying attention! Those who consider their message of vital interest are often disappointed!

Although a seemingly obsolete method, a printed newsletter that can be posted throughout the workspaces should be strongly considered with or without intranet communications. Again, are they read and taken seriously? E-learning and webinars can be a very effective method of communicating and training, particularly when distances between sites/facilities make it expensive and inefficient to gather all those who need specific information in a single location. Inserting program messages tailored to participants into classroom training and internal conferences uses valuable time most effectively.

To ensure that people will take time to absorb the message, communications must be concise and brief. Gaining attention requires a distinctive look that differentiates Operational Excellence-related messages from other facility communications.

> An organization implementing a major improvement initiative established an improvement program newsletter in a special format with a distinctive logo and colors. The logo: a butterfly visually represented metamorphosis from a caterpillar (the old organization)

into a beautiful new form. Symbolic musical notes in the left margins of communications were used to indicate an action that was "playing" from the enterprise improvement strategy.

Address all Anticipated Concerns

Many potential concerns are readily identifiable. Any proposed change, including reorganization, will be uncomfortable to many. As stated earlier, any mention of increased efficiency or reduced costs will immediately link to a fear of personnel reductions and/or reduced compensation as a result of factors such as diminished overtime. Many supervisors and managers will fear potential loss of position, authority, and status. All these concerns and more must be met head on as soon as possible.

If cost reductions are deemed necessary, personnel levels at objective costs must be determined early in the improvement process. This sensitive subject was covered in detail earlier in the chapter.

A highly experienced individual was recruited to improve an unprofitable facility with a woeful record of operating availability and cost. He developed and implemented an ambitious 2-year improvement plan. At the end of the 2 years, all objectives had been met, and availability, operating efficiency, and cost had been brought up above industry averages. The facility was profitable, and executives were very pleased with performance. His reward—he along with about half of his workforce were laid off as no longer necessary, as problems had been solved!

WHAT YOU SHOULD TAKE AWAY

Implementing Operational Excellence will likely require improvements to the organization. Improvements to elevate individual and operating effectiveness, encourage ownership and responsibility, facilitate the formation, function, and success of multidiscipline improvement action teams, maximize human performance, and move business value-oriented decision making lower in the organization.

Improvement begins with requirements: produce more, spend less, preferably expressed in terms of safety, environmental, production output, cost, etc. with measurable percentage improvement objectives. Communications are essential to create enthusiasm and support for improvements and Operational Excellence as the means. Messaging must be positive: why are we doing this, why is it necessary, and what are benefits to you? stressing the latter.

Values, relationships, and interactions between people are the starting point and foundation for the organizational structure. Operational Excellence demands a comprehensive set of values that assure commitment, honesty, trust, cooperation, personal, and organizational effectiveness.

Organizational improvement begins by identifying opportunities, strengths to build on, weaknesses to correct, and potential barriers to overcome. The ultimate objective is to move from management control to an effective team-based organization that safely and sustainably uses human resources most effectively.

A communications plan and team are essential to assure that all receive clear, consistent, and appropriate messages to inform and publicize successes and relieve concerns. Messaging must be configured to convey the right information to the right people. All methods of communication: electronic, intranet, printed, and others, must be considered and used to gain greatest exposure and understanding.

11

CONDUCT INITIAL TRAINING WORKSHOPS

Facilitated training workshops are a very effective means for conveying the details of Operational Excellence, as well as specific processes, practices, and procedures. They provide participants an opportunity to work through an abbreviated version with all the elements of the full process. A workshop to identify organizational improvements was addressed in Chapter 10. Similarly, a workshop can be employed quite effectively at the commencement of the Operational Excellence program to acquaint the Steering Team with concepts, principles, and the program itself. A workshop is an effective means to begin the process of establishing program requirements and developing mission and strategy mentioned in Chapter 9. A workshop is particularly useful to acquaint members of the Leadership Team with the details of the Operational Excellence program and provide a forum for refining the program plan, preliminary identification, and value prioritization of potential improvements. The latter workshop is ideally conducted when the Leadership Team is initially formed during the Define phase. The knowledge must be in place before commencement of the following, Identify phase.

Regardless of whether a workshop is to be conducted by an outside facilitator or internal resources, it should be preceded by an abbreviated assessment to ascertain conditions, strengths to build on, and weaknesses to be improved in the area to be explored. Strengths include people, their potential ownership, and commitment to an improvement program, processes, and programs in place and results. Weaknesses are potential limiting factors, including the organization and people. Experience has

Operational Excellence: Journey to Creating Sustainable Value, First Edition. John S. Mitchell.

indicated that a workshop is far more effective with existing conditions established and objectives defined, as it can be directed to specifics that are of maximum interest and utility to the Operational Excellence program, thereby creating excitement among participants.

The training workshop described in the following sections is primarily applicable to the Leadership Team. It has been delivered successfully on numerous occasions and is recommended at the initiation of an Operational Excellence program. The concept is equally applicable to workshops to introduce the Steering Team to Operational Excellence and begin the process of defining responsibilities for which they are accountable, as well as identifying opportunities for organizational improvement.

ESTABLISH TEAM TRAINING, FACILITATION, AND REVIEW PROCESS

Some thought should be given to the team process before the initial workshop. Where people are not accustomed to working in collaborative teams, team training might be considered before convening the initial workshop. Conducting ad hoc team training during the initial workshop may be more efficient and can be effectively directed at specific requirements identified during the process.

Ad hoc team training uses one or more trained facilitators circulating among the teams as they develop improvement opportunities. A good facilitator will immediately identify potential problems such as openness to ideas, arguing rather than discussing, one person dominating the team, as well as the opposite: one or more people who do not participate at all. In addition, the facilitator can identify discussions that stray from the subject or objectives of the workshop. Thus, the initial workshop becomes a learning process for both the Operational Excellence initiative and working effectively in teams. Facilitators should make a brief presentation at the end of closing presentations summarizing observations and things to work on in the team process.

An outcome of observations during the initial workshop might be the decision to conduct dedicated post-workshop team training based on observations and conclusions made during team interactions.

TECHNICAL TRAINING

Preliminary to, or as part of, the initial Operational Excellence workshop, some technical training in areas such as reliability, risk (Failure Modes and Effects Analysis), and methods including Pareto and Weibull analysis might be considered. Since participants will be expected to identify and analyze opportunities for improvement, it is imperative they have adequate working knowledge of the tools necessary.

WORKSHOP DESCRIPTION

The recommended workshop consists of 3–4 days, conducted in two mutually reinforcing phases. The first consists of a series of lectures where program and process are presented and explained in detail. U-shape seating is recommended to encourage questions and discussion that will ensure participants gain full familiarity with the operational excellence process. The second phase divides participants into small, four to six persons, facilitated work groups. Each work group develops one or two self-selected opportunities for improvement into improvement action plans using the Operational Excellence program sequence. Working groups present their improvement action plans to the whole group for comment on the final day of the workshop. It is highly advantageous to have a member(s) of the Steering Team and an enterprise/site executive or delegate present at the presentations to demonstrate support, gain a better understanding of the program, listen, and converse with the people who will have operational responsibility for program success.

The sponsoring executive of one improvement initiative considered the workshop and workshop results of sufficient importance to schedule his staff meeting to coincide with the closing presentations. In this way, his entire staff could participate. Recognizing opportunity and importance, the working groups spent most of the evening before the results meeting preparing very professional slide presentations. Many excellent questions were asked and candidly answered during the presentations, some quite surprising. The executive and his staff received a candid brief of perceptions at the working level that is often filtered out by the staff process. Participants saw executive participation as validation of the value of their efforts and departed highly motivated.

A very senior and highly respected corporate executive participated in the closing presentations of another workshop. Instead of working groups making a classroom-type presentation from the front, they arranged flip charts used in developing opportunities for improvement around the room. Each working group explained their opportunities, the process used to arrive at an improvement action plan, and considerations in detail. The executive stood with each group as the presentations were being made asking questions, making comments, and taking notes. Again, a valuable, highly motivating experience for all involved.

Workshop Process

Some of the program elements included in the workshop process will be covered in more detail later. The following outline and brief description are intended to provide an overview of a workshop that is considered essential to successfully launch the operating portion of the Operational Excellence program.

Presentation The first day and a half of the workshop typically consists of a slide presentation describing the necessity, benefits, and implementing details of the Operational Excellence program. As stated, it is best for participants to be seated in a U-shape rather than the more common classroom arrangement. The U-configuration is much more friendly for participants (who see response, realize others see their attention, and don't have to look at the back of a group of heads for 3 days!). The U-configuration facilitates attention, invites questions and comments, and produces much greater participation.

Working Groups At some point, either before the workshop is convened or during the first day, the individual organizing the workshop, usually the program leader, is asked to divide participants into working groups. Groups consist of four to six people who represent different functions in the organization and do not normally work together.

Identify Opportunities for Improvement Sometime during the first day, generally before lunch, participants are asked to list approximately five opportunities for improvement they consider of greatest value. There are no restrictions on areas or what improvements might be; determination of value is strictly up to the individual. Participants are requested to turn in their lists by the end of the day. Asking for opportunities for improvement before lunch typically stimulates and makes lunch and afternoon break discussions more animated and constructive.

With all opportunities for improvement submitted by participants and collected, the presenter/facilitator, sometimes assisted by the workshop organizer, compiles the opportunities into a single list. A spreadsheet has proven very effective due to its capabilities for editing and sorting. Experience indicates that considering duplication and overlap, a typical workshop of approximately 25 people will generate about twice that number of separate opportunities for improvement.

Group Development of Opportunities Working groups are typically assembled about 1–1.5 hours before lunch on the second day of the workshop. During this first group session, each group is asked to select two opportunities from the complied list to develop into an improvement action plan. To make every effort to have groups working on opportunities they consider of greatest priority, it is best to have each group select one opportunity in turn and then begin a second round in the reverse order, with the last group to select in the first round being the first to select in the second. With the working groups now having two opportunities to develop into action plans, participants are asked if they would like to switch groups to make certain that each group has maximum enthusiasm, ownership, and knowledge to work most effectively.

After lunch on the second day, working groups are allotted about 2 hours to begin development of the opportunities for improvement. All develop the opportunity in the following sequence:

- State the opportunity for improvement: may be modified from the original submission
- Describe the future state: what are the objectives, what will the process and system look like with the improvement fully implemented
- Define the value gained

- Describe the step-by-step process to achieve the improvement: the improvement action plan
- Identify strengths to build on, weaknesses to strengthen and challenges and barriers to overcome while implementing the improvement plan
- Define activity and results metrics (defined in Chapter 15) to measure progress and success with objective time lines
- List resources, people, and funding required
- Identify probability of success and define potential risks that might impede the plan's implementation.

Although all of the preceding elements have not been explained thus far, the workshop is arranged so that participants have the necessary information to develop and construct abbreviated improvement action plans. It is incumbent on the presenter of the workshop, presumably knowledgeable in all the program details, to circulate among the working groups to facilitate progress, answer questions, and make certain the work stays on track and discussions include all members of the group. This is the monitoring and facilitation to assure team effectiveness mentioned earlier.

Working groups are typically allocated the last hour of the second day to continue development of improvement opportunities. With this scheduling, participants frequently become sufficiently motivated to work until dinner and even into the evening. To encourage extended work, it is a good idea to hold the workshop in a location with minimal outside temptations.

On the basis of experience, opportunities for improvement during the introductory workshop are best developed on self-adhesive flip charts. This allows the group to iteratively develop the elements with the ability to see, modify, and add as appropriate. The flip charts are often used to make the presentation to the full workshop group.

To adequately develop two opportunities for improvement, the working group portion of the workshop should be allocated approximately 3–4 hours—1.5–2 hours for each opportunity. During visits to the teams, the facilitator must be conscious of time and make certain working groups don't spend all or most time on a single improvement opportunity. It is very easy for groups engaged in passionate discussions to lose track of time.

Value Principle Workshop participants typically experience greatest difficulty in valuing improvements. Many have considered improvement benefits exclusively in technical or operational terms—never in terms of value. This is a positive learning experience. It leads to emphasizing that people who have final approval authority for most improvement initiatives think only in value terms. Those developing the initiatives have to accomplish the translation from technical and operational benefits to business/mission value.

The learning process extends to the availability of data and information. Many will discover the data necessary to demonstrate value is either incomplete, suspect, or doesn't exist at all. This realization is often the beginning of a program initiative to improve the site/facility data and information system.

Group Presentations The workshop concludes with each working group presenting their improvement action plan to the entire group, preferably with executives and members of the Steering Team present to participate and ask questions. Approximately 15 min should be allocated for presenting each improvement opportunity. Since all initiatives were identified by participants, there is typically a great deal of interest and discussion. One of the biggest challenges for the facilitator is to encourage discussions while assuring sufficient time for all presentations.

The individual organizing the workshop, presumably the program leader, is encouraged to incorporate the key elements of each group's initiatives and findings into a workshop report that can be distributed, referred to, expanded, and refined as the full initiative is implemented.

REVIEW AND REFINE PROGRAM BASIS

After the initial workshop, the Leadership Team should review all program elements and refine as necessary. These will include program objectives, description, mission statement, policy, strategy, and expectations.

Although this task is more consistent with the Define phase, it is placed with the initial workshop, as the review requires insight and knowledge of the overall Operational Excellence process that will be gained from the workshop. As the program matures, the review and refine process become an integral part of continuous improvement. The program review and refine portion of continuous improvement essentially merge into the Define phase. Where the review and refine process finally ends up makes little difference so long as the Operational Excellence program itself is periodically evaluated, regularly refined and updated.

WHAT YOU SHOULD TAKE AWAY

A workshop is an excellent method to acquaint participants in the Operational Excellence program and especially the Leadership Team with Operational Excellence program methodology. Consisting of a mixture of lectures and interactive action team breakout sessions, a properly conducted workshop will provide participants with a working knowledge of the program and the procedures necessary to identify, value prioritize, and develop improvement initiatives.

12

IDENTIFY AND VALUE PRIORITIZE OPPORTUNITIES FOR IMPROVEMENT

At the conclusion of the Define process and introductory workshop, the Leadership Team will have a solid idea of program objectives and process. They will also have many ideas regarding initial opportunities for improvement that contribute greatest value to the enterprise business/mission objectives.

The first step of the Identify phase is to reexamine the Leadership Team and make certain that all the functions and disciplines necessary to implement the program are present and represented by active, committed team members. Reevaluating team composition, adjusting membership as indicated, is a continuous effort to assure all opportunities are represented with contributing people as the program moves forward.

With the Leadership Team solidified, opportunities to improve in areas that will contribute most to business/mission objectives and value are identified, refined, and value prioritized by contribution to enterprise business and mission objectives.

Many of the processes and examples cited in this chapter refer primarily to manufacturing/production enterprises, as these are more easily generalized. The same principles apply to mission-centric and administrative-type enterprises.

VALIDATE SCOPE OF IMPROVEMENTS

As illustrated in the summarized flow diagram shown in Figure 12.1, one of the first activities for the Leadership Team is to identify and validate all processes, systems, organization, and functions that will be affected by the Operational Excellence initiative. The identification process must assure boundaries are defined and contiguous

Operational Excellence: Journey to Creating Sustainable Value, First Edition. John S. Mitchell.

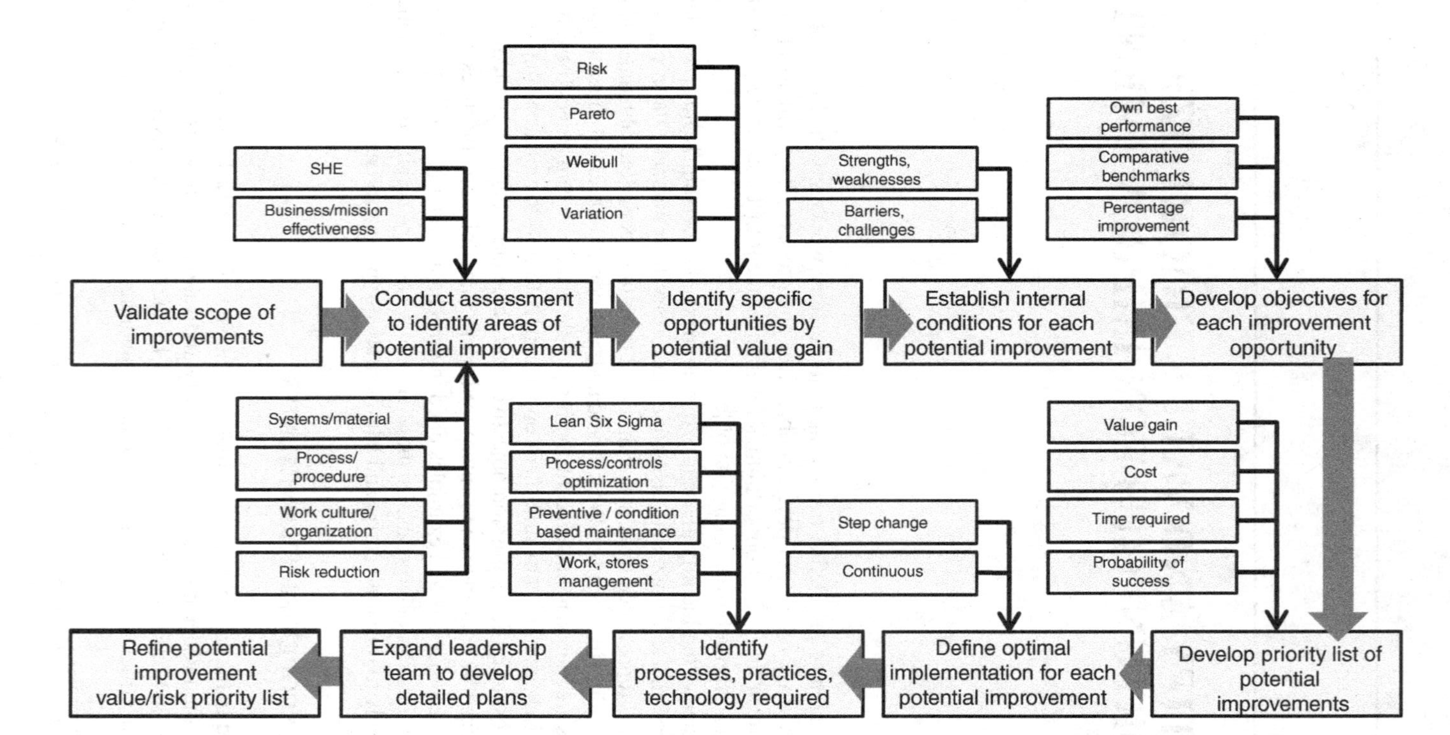

Figure 12.1 Summarized process flow diagram: identify opportunities for improvement.

and that all system elements, inputs and outputs, and the relationship between elements are fully and accurately described. Although this sounds like a task that should not require additional scrutiny, experience has indicated that in many operating facilities, process drawings, system boundaries, master equipment lists (MELs), and other vital documentation might not be up-to-date or accurately represent what actually exists.

> In one facility, it was discovered that a transfer conveyor between two process units wasn't considered owned by either. As a result, this vital mechanism received little or no attention—until it failed.
>
> In another facility, the master equipment list (MEL) was discovered so badly out of date, a task force had to be formed to verify nameplate data and change the MEL as necessary. The project required over 1 year to complete. In some cases, revisions approached 100%—nearly every entry in the MEL was incorrect!
>
> A facility that had implemented a less-than-successful Reliability-Centered Maintenance (RCM) initiative stated that forcing a clear, contiguous definition of systems and validation and correction of basic drawings and the master equipment list were highly valuable, unanticipated benefits of the process.

The same holds true for administrative improvement processes. Over the years, administrative systems are often informally modified by "dotted line" relationships that may be contrary to official organization charts but are essential to make the system work. "Improvements" made without recognizing how a system really works can create chaos.

> An electronics manufacturing enterprise initiated a project to improve the efficiency of a complex internal process. The process was directed to gather ideas for new instrumentation products from multiple sources and convert these ideas into detailed specifications for product development. After two intense weeks attempting to map the existing process flow, it became apparent that no single person fully understood the system then in use. Furthermore, the system had numerous instances of inefficient doubling back for reexamination and revaluation of features that had been approved. Ultimately, the improvement initiative was abandoned as "too difficult." The company was later forced into bankruptcy.

OPPORTUNITY IDENTIFICATION PROCESS

The opportunity identification and value prioritization process is directed to locating major departures from best performance. It proceeds in three stages as follows:

- Identify categories of performance and performers that are primary detractors from business/mission effectiveness
- Establish performance objectives
- Value prioritize improvements by probable gains achieved by elevating subaverage performance to objective performance.

Primary Detractors from Business/Mission Effectiveness

The opportunity identification process begins and is driven from prioritized categories of program business/mission performance: improve safety, health, and environment (SHE) performance, deliver more output, improve quality and efficiency, reduce downtime, costs, etc. Each is divided into specific subdivisions of improvement. For example, reduce costs can be subdivided into categories such as improve energy efficiency (reduce energy costs), minimize waste (downtime, scrap, necessity for rework, etc.), and reduce failures and maintenance costs. From here, specific processes, systems, and equipment with a history of poor performance are identified as potential opportunities for improvement.

Performance Objectives

It might appear counterintuitive to establish performance objectives as a second step in a process of identifying opportunities for improvement. Since Operational Excellence is an exception-based improvement program, identifying poor performers, systems, and components with historical problems is a logical beginning. Every operating unit/site/facility will have ample information identifying poor performers. If historical records are suspect, there will be a large amount of institutional knowledge regarding poor performance in virtually every area including the organization itself.

Performance objectives are established in at least two ways—actual best performance and published benchmarks. Performance objectives provide the "must be" basis for improvements from actual "where we are." The comparison of actual performance to the objective, often called *gap*, identifies the magnitude and potential value of a specific improvement; refer to Figure 12.2.

Large gaps in performance between "where we are" and "where we must be" may indicate the necessity for a step-change improvement described in the next section.

Actual Difference Between Best and Worst Performance The comparison method is to identify performance extremes within the facility/site/unit history. The difference, gap, between best and worst performance represents the opportunity for

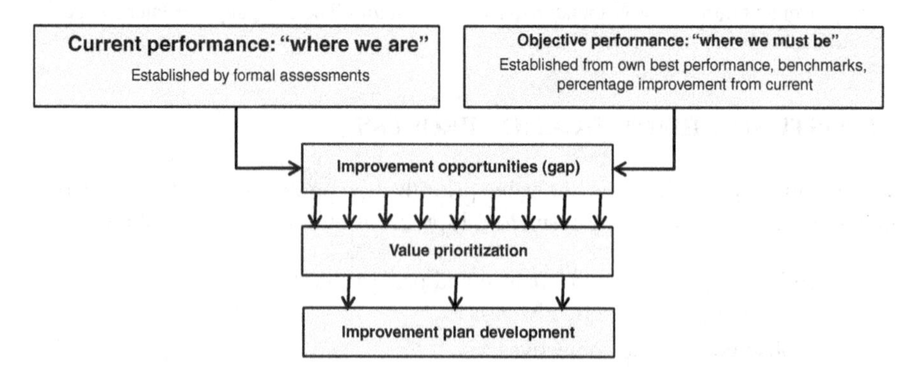

Figure 12.2 A comparison of current to objective performance identifies opportunities for improvement.

improvement. This method will be illustrated by segments on Pareto analysis following in this chapter, Pareto and Weibull analysis in the next.

The use of site/facility data as the basis for identifying and prioritizing potential improvements has significant advantages. It identifies actual poor performance with immediate meaning to program participants as deficiencies that have to be dealt with on a continuing basis. An initial focus on actual performance creates enthusiasm, ownership, and responsibility for improvement. Site/facility data should complement and validate internal anecdotal knowledge; if it doesn't, it is imperative to find out why!

Every organization has a vast pool of institutional knowledge of what works well, what doesn't, and ideas for improvement. Harnessing this knowledge, converting to improvement action plans, creates the ownership and enthusiasm necessary for success.

A key early challenge is to connect opportunities for improvement that generate enthusiasm and ownership at the working level into demonstrable value for the business/mission—the latter to generate credibility and support at the executive level. Sounds simple, but it's not!

Many improvement programs and activities that working-level people are certain create great value in terms of efficient operation; solving and avoiding problems are either not approved or later reduced in scope or terminated altogether. This often occurs because the true value perceived at the working level is not developed or communicated effectively as business/mission value and hence not considered worthwhile at higher executive levels.

Comparison to Benchmarks Best practice performance benchmarks are defined and published for numerous specific areas within Operational Excellence. Published benchmarks are an excellent reference to identify areas in which actual performance may not compare well with industry best.

Published benchmarks must be employed with some level of caution. In some cases, costs are one example, it is easy to conclude costs are excessive without a full understanding of why. Cost reductions without understanding why and taking proactive action to first eliminate the necessity for costs can result in increasing inefficiencies and ultimately increasing costs. Exactly the reverse of desired results! A second caution is there must be an accurate understanding of how the benchmark value was obtained and the method duplicated exactly in order to identify a real gap and opportunity for improvement.

An individual responsible for correlating performance metrics of individual business units (BUs) within a multisite operating enterprise noted that annual rankings of performance and year-to-year improvements weren't always consistent and didn't agree with estimates. His conclusion: some BUs were altering calculation methods to make their performance appear better. As just one example, availability expressed as a percentage can be greatly influenced by the objective. By removing periods such as downtime for changeovers and scheduled maintenance from the availability objective, a significant amount of nonavailability can be hidden.

Another individual commented that his enterprise used an overall effectiveness metric for internal comparison. On closer examination, it was recognized that each site was using a different method of calculation. His conclusion was to use an industry benchmark as a starting point and then measure improved performance as a percentage gain.

IMPROVEMENT METHODS

Safe, sustainable improvement is not a command but rather the result of doing the right things.

There are two methods of improvement: step-change and continuous improvement. In a real-world operating environment, both must be considered on the basis of conditions. Step change when the gap, magnitude of the improvement opportunity, is large, continuous when the gap is small. In a large, comprehensive improvement program such as Operational Excellence, both are likely to occur simultaneously on different initiatives. In most cases, a successful step-change improvement will transition into continuous improvement when the initial objectives have been achieved. In some cases, continuous improvement may prove insufficient for rapid external changes in the operating/market environment. In this case, it will be necessary to shift to a step-change process.

Improvement is not an event but a process. Gaining and sustaining results from improvements require continuing attention and effort.

Step-Change Improvement

Many advocate step change as the only way to achieve objectives quickly enough to reach competitive business/mission performance in a changing environment. As introduced in Chapter 2, step-change improvement assumes the future state can be defined completely and accurately. This may be true for certain business/mission elements and work processes but may not be true for others. When it can be accomplished, step-change improvement has the following attributes:

- Large opportunities for major change identified throughout the entire enterprise and site/facility to include organizational structure, roles, responsibilities, values, culture, methods, processes, and procedures
- Major opportunity for breakthrough improvement in performance and effectiveness and large leap into an ideal future state
- High risk/high potential gain strategy: creates greatest value most rapidly and when successful, is very rewarding organizationally and personally
- Many changes may be carried out simultaneously; the challenge of coordination and management must be considered
- Transition plan all important to avoid costly interruptions as major changes are implemented.

A large organization operating with seemingly large gaps to optimal performance and effectiveness was mandated to improve. An objective of 12% improvement in 5 years was proposed. Such an unambitious objective immediately moved all affected into a task-protective role on the supposition that the objective could be satisfied by minor alterations without addressing any of the gaps that demanded major improvement.

Continuous Improvement

Continuous improvement is advisable in areas where performance and effectiveness are reasonably close to best practice expectations as determined by a formal assessment; refer to Chapter 17. This includes SHE compliance, business/mission effectiveness, performance to enterprise and site/facility objectives, effective organization, and full implementation of best practice cultural values, processes, procedures, and methods.

Characteristics of continuous improvement include the following:

- Builds from current structure, policies, and procedures
- Changes are incremental and designed to optimize existing practices and procedures
- Ensures little disruption and an orderly transition
- Low risk/medium gain.

A continuous improvement process that sets annual objectives is critical. Knowing when to say "no" to more ideas is the difficult part. It is easy to want to accomplish all the good ideas, but there are always capacity limits to perform the work. It is important to select an achievable volume of work and perform it to successful completion—"if you chase too many rabbits you don't catch any"

IDENTIFYING SPECIFIC POTENTIAL IMPROVEMENTS

Identifying opportunities for improvement through a series of workshops was described in more detail in Chapter 11. As opportunities are identified, improvement action teams are formed to fully define the opportunity, results/end state to be achieved, and to develop detailed plans to move from the current to the improved. The team members selected to address specific opportunities are typically one or more from the Leadership Team having special interest and knowledge and willing to champion and lead the initiative. The team is augmented by people with specific knowledge, who wish to participate and can contribute.

The team-based improvement identification process typically consists of the following:

- Appoint a top-level coordinator. This is a requirement when multiple teams are chartered to address one or more interrelated improvements. This individual must participate in all teams. The coordinator is authorized to adjust team charters and objectives if and when necessary to ensure consistency and mutual support in the improvement process. Furthermore, to assure all improvement initiatives work together effectively to produce an optimum result.

- Team composition is determined by area and opportunity. All interests and institutional knowledge of the particular area/opportunity must be represented within each team
- Establish team rules and conduct team and technical training as required. Essential attributes for the team include common objectives, trust, effective communications, courtesy, and cohesiveness. It is also necessary to establish methods to resolve any potential friction and conflicts within the team. Technical training in processes and tools to be used during the process is conducted as required. *Conflict during an idea-generating session is typically one of the first disruptions experienced by a new team. When this occurs, everyone must be reminded that idea-generating sessions (brainstorming) are not the venue for detailed discussions of the merits or potential flaws of a given idea. Rather, the objective is directed solely to identifying ideas. Discussions of the pros and cons, benefits, and potential barriers to success of individual ideas occur later during a feasibility and value prioritization process, after as many ideas as can be identified are listed, and each is opened for detailed examination and discussion.*
- Define the direction, opportunities to be addressed, and preliminary conclusions from the initial evaluation. Strengths to be preserved and boundaries are all considered and established at the outset of the process. Additional topics for discussion include objectives, results required, potential reorganization including roles, responsibilities in a revised organization, benefits, and any expected difficulties or barriers. An initial transition plan from current to new and the time allowed to accomplish the results are included. Issues must be completely identified and discussed openly. There must be good participation from all members. *Every team appears to have one or more people who tend to take charge and dominate discussions (often not the person with the most to contribute) and one or more who are content to remain silent (often an individual with the most to contribute). An experienced facilitator will recognize this dynamic and take steps to assure effective participation and contribution from all.*
- Each team meeting must be preceded by a detailed written agenda listing subjects to be discussed, objectives, responsibility, and participants. Meetings must be followed with a written report, including highlights of the discussions, actions taken, and actions required.
  Because of the complexity of discussions and the difficulty of simultaneously taking notes and participating, especially for the facilitator, it has been found helpful to make a voice recording of entire meetings. The recording can be reviewed later to assure all nuances of complex, often heated, discussions are accurately reflected in the written minutes. Experience has demonstrated that recording meeting discussions has no adverse impact on the openness of discussion by participants.

- Build ownership, commitment, motivation, enthusiasm, involvement, mutual respect (members respect and support one another), and cooperation.
  By its nature, the team process builds the initiative, ownership, and teamwork values that are required in the new organization. In addition to using institutional knowledge most effectively to develop an improvement plan, team development builds positive relationships with others involved in the activity and plants the seeds necessary for full cooperation and ultimate success.
  Recognize that it will take time, in some cases years, when a major shift in institutional culture is required, to get people aligned and working together effectively in teams.
- Identify actions required to achieve the new objectives. These may include redefining processes, roles and responsibilities, additional resources, technology, and training. Practical action plans are formulated to meet objectives including resources required, risk evaluated, value calculated, and measures of performance
- Decisions made by consensus
  No matter how hard everyone may try, it is often impossible to go beyond 70% or so consensus. When that occurs, all must agree to fully support the majority, with the proviso that if results later demonstrate the 30% way was best, the plan will be adjusted. Leadership is crucial to assure this happens. Refer to Chapter 10.

It must be noted that gaining participation of the best people at an enterprise, site, or facility for an improvement initiative will likely require some schedule flexibility. Since the people who can contribute most are also likely the busiest with heavy responsibilities, it is generally best to stagger schedules so that people participate in the Operational Excellence initiative a week or so a month, staggered 2–3 days at a time. Also recognize that the site/facility may have its own ideas of when people can be made available and that interruptions and rescheduling cannot always be avoided. Flexibility is essential.

QUANTIFY AND PRIORITIZE OPPORTUNITIES—IN BUSINESS TERMS

Once poor performers are identified, site/facility best performance can be used as performance objectives, as well as industry benchmarks. Both must be considered. In many cases, an interim objective of elevating worst performance to population average will represent a major improvement that will be enthusiastically embraced by team members.

> An oil refinery with approximately 2000 centrifugal pumps calculated an average 40-month Mean Time Between Failure (MTBF) for the population as a whole. The MTBF for the 10% poorest performers was 12 months; average repair cost was $5000. Doubling MTBF for the poorest performers to half the population average will result

in an annual savings of $5 million—a clear opportunity for improvement. With the opportunity clearly defined, specific improvement activities and tasks can be effectively concentrated on the small population of poor performers to assure rapid results.

Teams are typically very good at identifying opportunities for improvement, but they often need help determining the value of improvements that will gain attention and be attractive to executives.

A value model and process similar to that outlined in Chapter 5 is useful to establish a valid comparison of cost and value gained by improvement initiatives to increase availability and effectiveness and reduce spending. Results gained through the value or a similar model will be credible to finance, business, and financial executives. With a broader knowledge of the value creation process, financial elements of the business and the balance between increased production and reduced spending, the Steering Team may have to help during the early stages of establishing value for prioritizing improvement opportunities.

Value Prioritization

In each case, the objective of this analysis is to identify processes, systems, and individual equipment that constitute the greatest deviations from optimal performance. The next step is to identify reasonable results that should be expected from improvements for the purpose of value prioritizing implementation. Value prioritization is based on the cost to implement, the risk, and value returned. Value returned includes contribution to program objectives that align with and support enterprise mission/business objectives. Whether the primary objectives are to improve mission compliance, production output, reduce costs, some combination, or additional considerations, accurate data is the essential starting point.

Necessity for Complete, Accurate Data

Accurate data is essential to identify the source/cause and magnitude of production/mission losses, downtime, failures, costs, waste, and inefficiencies of all types from which to develop potential improvements that increase the delivery of business/mission value. Furthermore, once opportunities for improvement have been identified, accurate data is required for prioritization to assure improvement initiatives are implemented in a sequence that makes greatest contribution to increasing business/mission value.

It often occurs that people being introduced to Operational Excellence quickly recognize that accurate data is a formidable challenge, particularly in the value prioritization phase. The existing data structure may not provide an accurate, easily accessed list of deviations from objective performance. Deviations may be quite tedious to reconstruct from operating records.

The difficulty can be lessened by initial screening to identify areas where the time necessary to research and create accurate defining data will have greatest value. A Pareto analysis, described in detail later in this chapter and again in Chapter 13, is a highly useful tool to quickly identify the greatest opportunities within a large data set so that improvement efforts can be concentrated in areas of greatest potential value.

Once the necessity for accurate data is recognized, creating/improving a comprehensive data gathering, storage, and access system is a logical next step. If data recorded previous to program initiation are considered suspect, compiling an accurate history of events and work accomplished must be commenced immediately. Accurate history begins at program day one!

> Participants in a major, site-wide improvement program recognized the need for comprehensive data about 18 months after commencement. At that point, revising procedures that had been in place only a short time, adding new procedures and retraining personnel who were just becoming comfortable in "the new way," proved a formidable task that was resisted by many. Lesson: make certain provisions for all anticipated requirements—especially accurate data and information, are incorporated in the Operational Excellence program from the beginning.

Identify Below-Average Performance and Performers

As stated, it is highly advisable to begin the search for improvements in areas that are important and familiar to the Leadership Team—particularly in the early stages of an Operational Excellence program.

It bears repeating that beginning the program with emphasis on identifying and resolving poor performance identified by team members immediately creates enthusiasm and ownership. It demonstrates the Operational Excellence program will improve conditions those at the working level must deal with and resolve on a regular basis. Commencing with recognized, immediate opportunities for improvement is far better than beginning with more distant or generalized opportunities. Reducing failures and downtime are two areas likely to generate immediate interest. This aspect of an Operational Excellence program will be revisited later.

History

Within a typical operating enterprise, operating and maintenance histories are logical starting points to identify opportunities for improvement. If reducing downtime is one principal objective of the Operational Excellence initiative, the search for improvement opportunities begins with a listing of all downtime by cause and cost over a given period, typically 12–18 months. (Note: total operating and maintenance cost may not be immediately available; elapsed time generally is.) This assumes that without action, poor performance in the past is a likely precursor to future performance. This concept will be expanded later in this chapter. Similarly, a star performer in the past will continue to be a star. There is a bit of a nuance; as will be illustrated later and occurs in many endeavors, there may be categories, systems, and assets with good performance in the past that are trending toward lower performance. These need to be identified.

CATEGORIES OF PERFORMANCE

Following are a few categories of performance to be analyzed to identify detractors from best performance. With discrepancies defined, improvement initiatives will be valued, developed, and implemented by contribution to business/mission effectiveness.

Safety Health Environment

Identifying discrepancies in SHE performance from which to develop opportunities for improvement follows a defined process. A search begins from the site/facility records of SHE performance exceptions: incidents and near misses. As these records are a legal requirement in most areas of the world, nonexistence is a major deficiency that must be corrected as soon as possible. SHE incident and near miss records are screened to identify major, costly incidents by date and time. Later in the process, an effort will be made to link SHE incidents to production and other operational losses, as production disruptions sometimes lead to decisions that are not fully thought through and cascade into other areas. Several highly publicized events within the last few years attest to how this can happen and the importance considering risk in any decision process.

Conformance to Business/Mission Objectives

This area typically consists of mission compliance and production effectiveness, achieving consistently greater outputand reducing variation, failures, and costs.

Records should be available or constructible to identify departures from best-attained operational performance for further analysis. Weibull analysis, described in detail in the next chapter, is a very powerful, particularly useful, tool for identifying specific deficiencies that detract from production output.

Downtime There must be a solid definition of downtime that will separate time out of service due to a failure or similar occurrence from downtime because the system or equipment is not required or is in standby. For the purposes of Operational Excellence, identifying substandard performance for improvement, it is imperative to separate a failed mode from in service but not operating. ISO14224 provides general guidance; refer to Figure 12.3.

Costs Costs are analyzed to identify and rank processes, systems, and equipment that are above-average consumers. When the population of similar consumers is large, similar equipment as an example, the Pareto analysis described in detail later in this chapter is highly useful to identify specific members of the population with substandard performance for more detailed analysis. Costs should be available from the enterprise business system as a basis for Pareto analysis to identify primary anomalies by system and equipment asset.

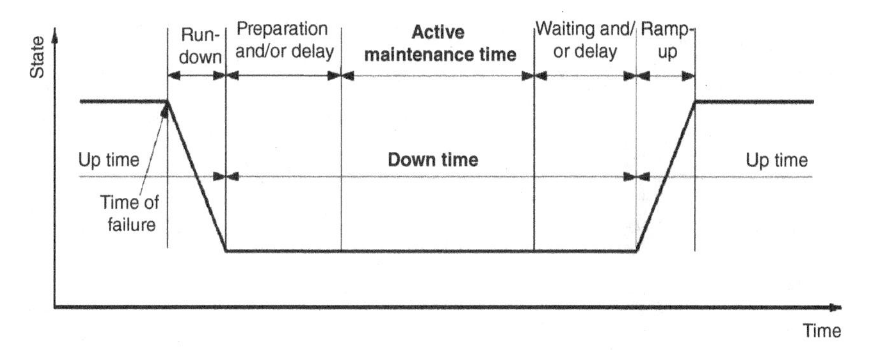

Figure 12.3 Operating phases, from ISO14224.

Production costs must be recorded with sufficient granularity to permit accurate accounting. Utility costs such as electricity, gas, air, and steam should be metered by unit to allow accurate allocation, tracking of improvements, and full credit for improvements in efficiency.

Multiple Objectives

The preceding examples assume there is a single improvement objective, for example, increase mission compliance, production output, or reduce failures and cost. A Pareto analysis can identify the greatest potential contributor to a single objective. But how to identify opportunities to create greatest value when there are two or more concurrent objectives, for example, increase production output *and* reduce cost? Figure 12.6, later, will illustrate one method.

METHODS FOR IDENTIFYING OPPORTUNITIES

Chapter 13 will describe Weibull analysis using daily production data to identify and prioritize opportunities for improvement. The Weibull graphical presentation is a clear, understandable, and convenient method to identify deviations using actual performance data. Once identified, deviations are opportunities to identify and eliminate cause. Eliminating the cause of deficiencies improves operating processes and leads directly to increasing the success and profitability/mission effectiveness of the enterprise to the benefit of shareholders and employees alike.

Pareto Analysis

The purpose of a Pareto analysis is to identify the greatest contributors to a given outcome among a (typically large) data set. A Pareto distribution states that 80% of a given characteristic will be located in approximately 20% of the total data set. Originally devised to describe the distribution of wealth within a population (80% of

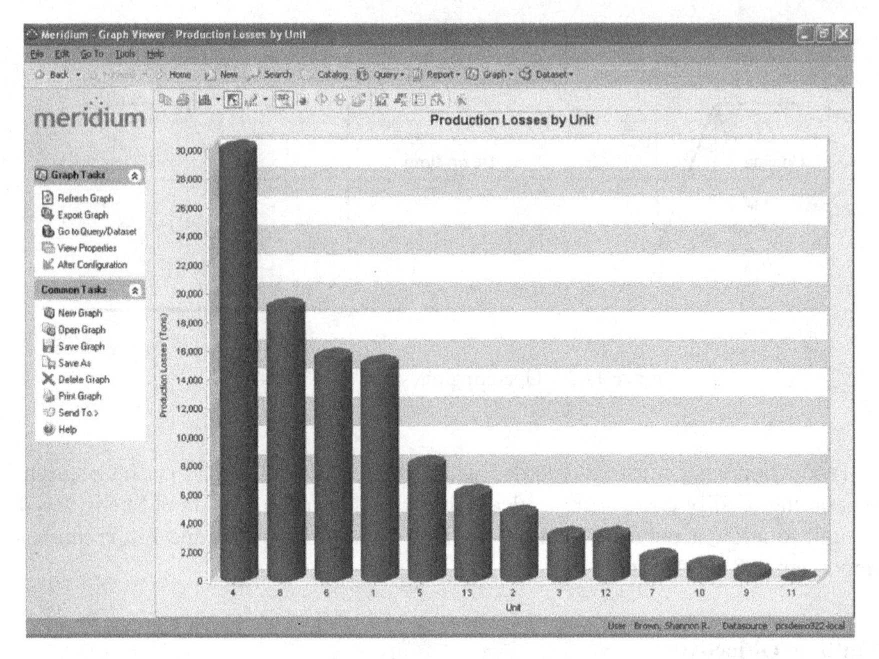

Figure 12.4 Pareto analysis identifies production losses by operating unit.

wealth held by 20% of the population), a Pareto distribution is often used to identify major deviations from objective performance. As an example, within an operating enterprise, a Pareto distribution states that 80% of total losses will be attributed to 20% of the total events. A Pareto analysis quickly separates a complex data set by importance, allowing time and resources to be concentrated into areas that potentially yield greatest results.

Graphical Pareto analyses are easily accomplished with a spreadsheet program. Many specialized applications also include provisions for Pareto analysis with added features.

Figure 12.4 illustrates a Pareto analysis of production losses by unit within an operating cascade of 13 units. It was produced from a spreadsheet of daily production losses over an 18-month period compiled from operating logs.

The Pareto graph in Figure 12.4 shows that 4 of 13 units (4, 8, 6, and 1) are responsible for over 95% of the total losses. Improvement efforts with greatest potential impact on production are clearly concentrated most effectively on the first two: units 4 and 8. The next step is to determine cause of production losses within the two units: a step that was accomplished by a second Pareto analysis.

Figure 12.5 is the Pareto analysis of production losses in units 4 and 8 by cause from the same data.

It is interesting to note that the two highest causes in Figure 12.5 are “Unknown” and “Other.” This is not an unusual outcome with real data recorded with little appreciation for use.

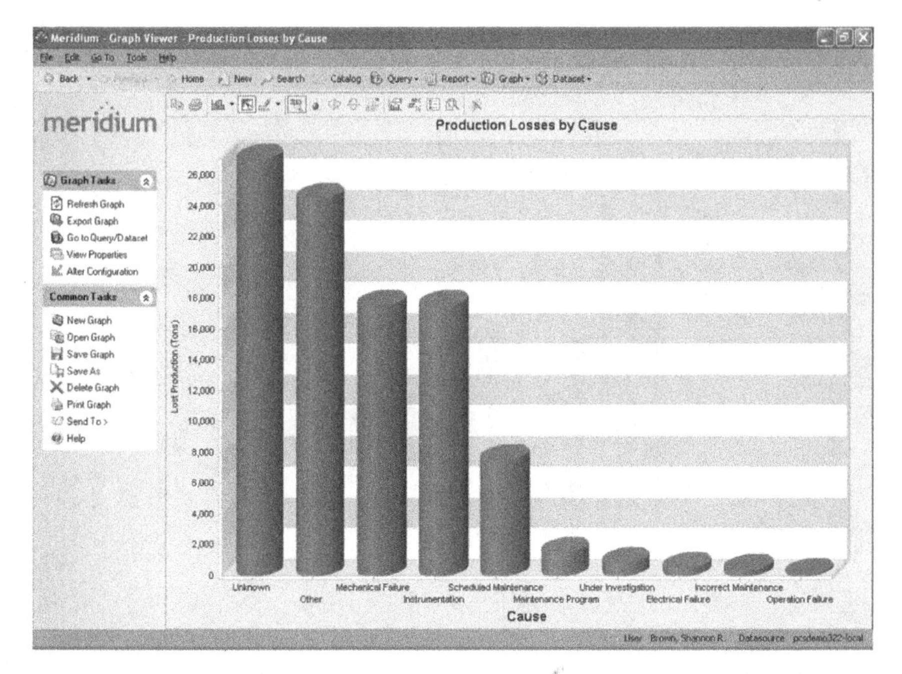

Figure 12.5 Pareto analysis of production losses by cause.

In this specific case, a third Pareto analysis was used to rank individual production loss incidents within the unknown and other categories by date and time in order of duration (magnitude) of loss. With the data set of primary interest narrowed to approximately 20% of the total loss incidents, the effort to research and identify cause from operating logs and other documentation became very manageable. The next step was to redo the data transferring large losses, now defined by cause rather than unknown and other, into a defined cause category until the unknown and other categories had been reduced to 20% or less of total losses.

One of the significant losses was quite interesting. A worker waved to a friend driving a fork lift. Waving back the forklift driver's attention momentarily lapsed with the result that the extended lift contacted and broke off a vital instrument halting production for a significant time. Records attributed the loss to instrumentation. But was this really the cause? A good case can be made for a design error that placed a vital transmitter in a vulnerable location.

As will be emphasized in the next chapter, care must be taken to assure a Pareto analysis represents total impact. The Pareto analysis shown in Figure 12.6 and mentioned earlier displays both cost and production loss on a single Pareto chart.

Note in this graphic the cost of lost production (left bars on each entry listed along the horizontal axis) far outweighs maintenance cost (right bars). Based solely on reducing maintenance costs, improvement efforts would be concentrated on Pump

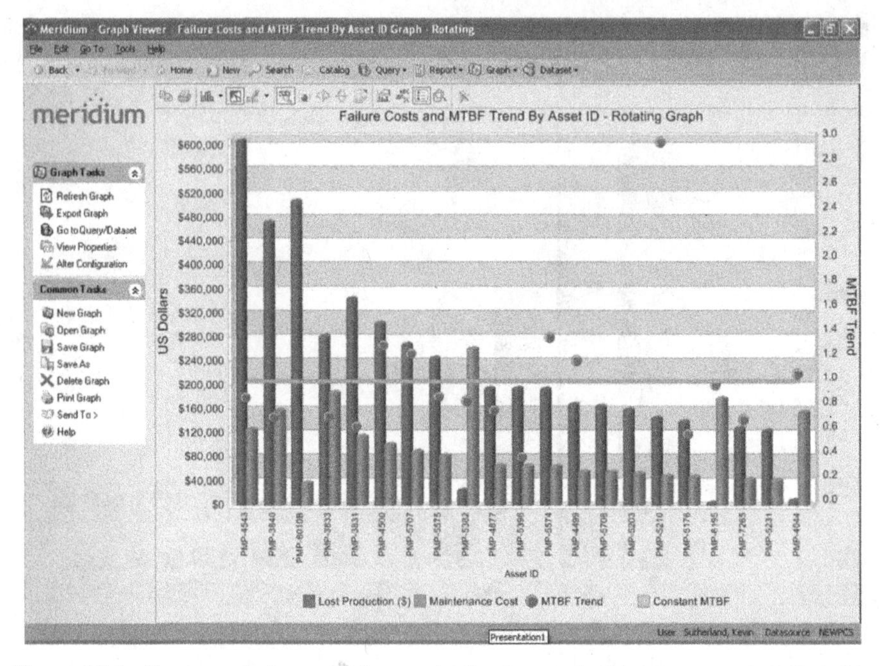

Figure 12.6 Pareto analysis combining production loss and maintenance cost attributed to specific equipment assets.

5382 in the center of the graph. When the operating and maintenance costs are combined into total cost, the same pump doesn't rank in the top 10 improvement opportunities. In this plant, maintenance costs on a specific pump are, on average, less than half of production losses. It should be noted that this particular application program contains a third, very powerful, bit of data: a normalized MTBF trend denoted by filled circles. One of the pumps, 5210 in the right quarter of the graph, is trending positive: a very good sign that indicates costs are declining. Others, including 5396 in the center, are trending negative (MTBF decreasing). Not a good sign!

The real issue in creating this presentation is assuring the financial benefits available through increasing production (reducing loss) and reducing cost are normalized to equivalents. That will be addressed later in this chapter.

Additional Considerations with a Pareto Analysis Within a production/manufacturing environment, experience has indicated that instead of the conventional 80-20 distribution, problems and high costs may well be closer to a 90-10 distribution: 90% of the problems and high costs attributed to 10% of the equipment population.

A further consideration: within the maintenance area, the opportunity analysis can be extended to factors such as the following:

- Origin of emergency and schedule break work. This is important as unexpected work that requires immediate changes to the established work schedule is less efficient and causes widespread disruptions that must be fully considered in a value prioritization.

- Repair success: work quality and potential design defects where repairs are less than successful. Investigation in this area may lead to the conclusion of a shortcoming in design or operation, need for a more detailed operating, or repair procedure or broader, greater depth training.

Distribution within the analysis will provide accurate guidance for improvement opportunities.

RISK

To open with a disclaimer, the following is a very brief discussion of an exceptionally important topic. Risk identification, analysis, and mitigation are areas in which knowledge, adherence to procedures, excellence, and compliance to regulations and standards are imperative.

Analyses in the preceding section were all based on historical performance on the assumption that past performance is likely to repeat in the future. Operational Excellence turns this assumption on its head. Operational Excellence is directed to identifying opportunities for improvement so that the future will be better than the past. Hence, the necessity to recalibrate opportunities from time to time and continue to look into the future. As improvements are implemented, performance gets better.

For a complete opportunity analysis, it is necessary to estimate the future—risk. Risk, defined as probability times consequences considers threat, vulnerability, and impact of an event.

A robust risk identification, assessment, ranking, and mitigation/control procedure must include as a minimum:

- Risk calculation methodology in accordance with a recognized standard
- Risk rank matrix of systems and equipment
- Detailed procedure for mitigating controls for highest risk systems and equipment
- Requirements for use of Failure Modes, Effects Analysis (FMEA) or an equivalent
- Requirements for risk consideration in all decisions.

As illustrated in Figure 12.7, probability is the likelihood of an event ranging from very rare, but could happen, to chronic, happens all the time. Consequences describe what happens if the event occurs including impact on SHE, business interruption, lost production, reduced quality, operating and restoration cost, waste, and fines. Consequences range from minor to avoidance imperative. The latter is typically a serious SHE incident and/or large production loss/business/mission interruption. Various

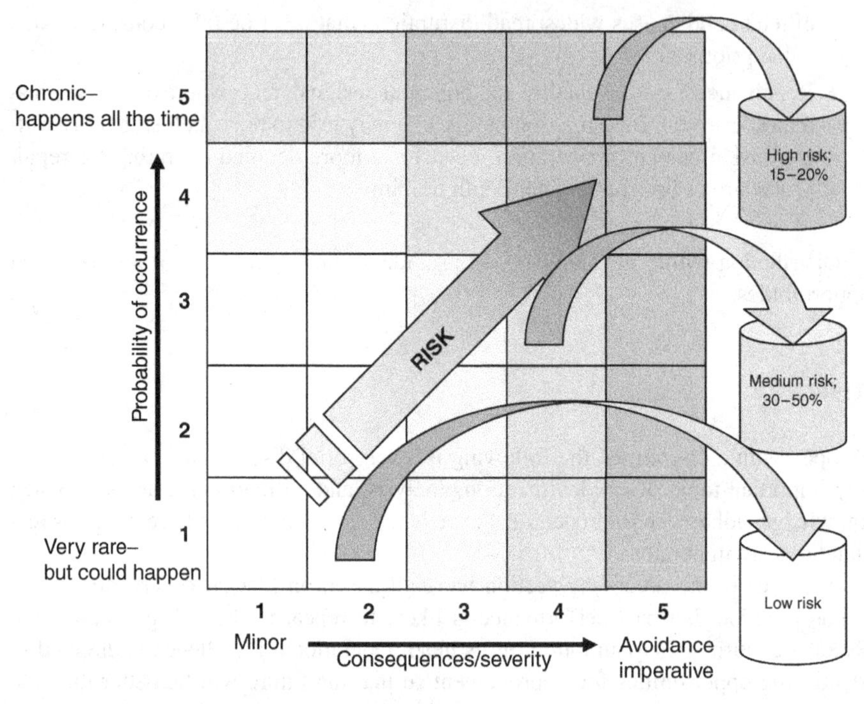

Figure 12.7 Risk matrix.

methodologies place numerical values on both probability and consequences. Generally speaking, once an event is identified, it is relatively easy to define consequences. Estimating probability is quite difficult.

A thorough risk analysis assures attention and resources are applied most effectively to optimal operations, surveillance, fault detection, improvements to systems, and components with highest risk. The goal is to proactively prevent failure rather than concentrate on response.

There is an important factor to consider in risk and risk analysis. Within many production/manufacturing facilities, criticality and risk are used synonymously. Problem arises when criticality is interpreted as consequences; a criticality analysis is conducted without full or any consideration of probability. When this occurs, most systems and equipment will be classified "critical." The prioritization achieved with a full risk analysis considering probability *and* consequences is effectively nullified. Everything is high priority. There is no prioritization for surveillance, work requirements, etc. The resources necessary to assure maximum risk mitigation may be spread too thin to assure optimal control.

At the beginning of a major improvement program, a site classified systems and assets by criticality to establish a priority for applying improvements. Of approximately 2200 systems, about 1800 were classified as highly critical. That division provided absolutely no guidance as to where initial efforts should be applied to achieve greatest risk reduction. The program eventually failed; lack of prioritization was one of the primary reasons.

To make certain there is a valid priority based on risk, it is generally necessary to require the consideration of both probability and consequences and stipulate that no more than a stated percentage of equipment and systems can be classified in the highest risk category. This forces the risk analysis to fully consider both probability and consequences. A good starting point is 20%. Best-in-class performers refine the risk analysis even more to assure attention and resources are concentrated where they are needed most and have greatest value.

Risk Analysis

A risk analysis is a formal, systemized process to anticipate the future, including the means and likelihood of detecting incipient symptoms in time to prevent or minimize the occurrence and/or consequences of an actual event. Risk analyses are accomplished on piping and pressure vessels, mechanical and electrical systems, instrumentation, fixed, and rotating mechanical equipment, structures, and more. In enterprises with hazardous processes, comprehensive risk analyses are a requirement.

During a discussion about determining risk, an employee of a large, global company was asked how they assessed risk when the potential consequences might include one or more fatalities. The reply was that the enterprise had a thorough, realistic procedure for risk assessment and control. The probability of a fatality was treated with the same detailed engineering logic and evaluation used to analyze all potential high consequence events. In his opinion, it was far better to use consistently best engineering judgment rather than treat potential fatalities differently because of probable litigation.

There are numerous specifications for risk analysis, management, and control published by ISO (ISO31000), American Petroleum Institute (API), American Nuclear Society, and others. The process is amplified by detailed instructive material. Every enterprise must have a risk identification, analysis, and management/control procedure in place that is in full compliance with one or more internationally recognized standards for the specific industry.

Similar to the opportunity analysis in Operational Excellence, risk is addressed in a top-down priority sequence where the risks posing greatest threat; combined probability of occurrence and *consequences* (impact or loss) are evaluated and mitigated first. Risks with lower combined probability and consequences are addressed in descending order.

Figure 12.7 appears quite simple; however, the apparent simplicity, typical color coding, and numerical values are very deceiving. The actual process of assessing overall risk, specifically probabilities mentioned earlier, is highly challenging. A typical enterprise can't coexist with events in the upper right hand corner. These demand correction—typically with layered controls to reduce probability and/or consequences. Control actions in several categories must be taken to reduce risk to an acceptable level.

Passive: generally includes design elements implemented to primarily reduce consequences. These include many familiar methods: limited access highways, physical barriers between opposing highway lanes, automobile crash resistance including air bags, standards for home wiring, flame resistant upholstery, guards around power tools, and fire prevention barriers. Industrial risk mitigating measures include restricting access, containment for power generating nuclear reactors, locked fencing around high voltage electrical distribution equipment, placing control rooms, and other normally occupied facilities outside operating units and other similar measures. Some of the preceding examples reduce both probability and consequences.

Active: includes features to recognize and interrupt/mitigate an event (probability and consequences) such as safety and pressure relief valves, Safety Instrumented Systems (SIS), warning alarms (including building fire and CO), automatic trips (electric circuit breakers, overpressure, overspeed, elevator doors, etc.), and manual trips such as found on escalators.

Procedural: typically avoidance (reduce probability) including procedures for starting and shutdown of an industrial facility, lock out, tag out, closed area entry and work, working aloft (including safety instructions for working on ladders: probability and consequences), hot work, multiple pressure barriers, etc.

A typical risk avoidance scheme will use measures in all the three listed categories to reduce and control potentially high risk events.

Challenge occurs when decisions, made without fully considering risk, nullify one or more control measures without fully considering the added risk being incurred. Referring back to Figure 12.7, several control measures may be in place to reduce risk from an unacceptable location in the upper right to somewhere in the middle to lower right. There have been recent events where people, typically under pressures of time, make decisions that abrogate one or more barriers and thereby increase risk—seemingly without a full assessment of additional risk incurred, with disastrous consequences.

It is absolutely imperative to thoroughly understand the concept of risk depicted in Figure 12.7 and thoroughly evaluate any decision that could degrade the risk profile before taking action.

Two final comments about Figure 12.7: first, “Black Swans” live in the lower right hand corner. In this area, there may be one or more potential events identified with massive consequences yet an infinitesimal probability of occurrence. In an analysis with strictly numerical criteria, this potential event would likely fall below the horizon of risks to be considered. In actual fact, “Black Swans” must be included in a complete risk assessment, thoroughly analyzed and treated with the same seriousness as risks with comparable consequences and higher probability. Action must be developed and implemented for earliest recognition of incipient symptoms and full mitigation/control of consequences.

Regarding the Fukushima Daiichi nuclear power plant incident, it is quite clear (admittedly in hindsight) that a thorough risk analysis should have considered consequences of breeching the seawall—no matter how unlikely that appeared to be. Had a breech of the seawall been considered, it is very possible that mitigating action would have included moving emergency generators and their fuel source to higher ground, routing emergency power overhead, assuring power to reactor coolant pumps and a host of other alterations.

The second area of interest, the upper left corner of the risk matrix, represents an area where excellence and value collide. Corrective actions are often ignored or never implemented for chronic problems that have little or no consequences. This can create a level of cynicism at the working level. "How can an enterprise say they are totally committed to excellence and, at the same time, allow nagging, repetitive problems to remain?" The answer is they must be corrected; the commitment to excellence must override value in situations such as this.

Reliability engineers at a large facility committed to excellence in all aspects of their operation continually complained of small, but aggravating deficiencies that couldn't be approved for correction due to low value. An option presented for consideration was to allocate a small budget, approximately $50,000, to each reliability engineer that could be spent for improvements that did not cross the value threshold for normal corrective work. The proposal reaffirmed the facility's commitment to excellence and empowered experienced, committed, and responsible people to determine corrective priorities. A win-win for Operational Excellence.

Future operational requirements must also be considered during the risk assessment. These include any known future changes in operations or the operating environment that could affect value produced by the mission/business. An example might be changes in raw materials or mission requirements that affect operating intensity. Changes that would cause systems and equipment to be operated harder and/or in a more severe environment. There may also be potential changes to the regulatory environment or external threats to the business such as lower cost competition, changing market or supply conditions. Finally, has the site/facility been operated in an unsustainable manner that will require investment to make up for uncorrected deterioration?

An industry-leading facility, very aware that recurring budget constraints in a highly competitive cyclical business resulted in unsustainable operation in many areas, developed processes to estimate the cost of deferrals on the basis of net present value. This not only provided excellent guidance for prioritization, but was also the basis for justifying investment for restoration when the budgetary crisis had passed.

Failure Modes and Effects Analysis

FMEA from RCM is an excellent process and starting point for risk analysis. Illustrated in Figure 12.8, a typical FMEA template includes a description of the anomaly (Failure), component and cause (Mode), Effects (consequences), probability, likelihood of detection, and methods of control/mitigation. The likelihood of detection warrants additional explanation. It includes how a failure is recognized. Is a potential failure immediately apparent; is there a warning interval before actual failure?

Failure Modes and Effects Analysis (FMEA) Template

System: ______ Date Created: ______ Site: ______ Team Members: ______
Sub-System: ______ Date Revised: ______ FMEA Facilitator: ______
Equipment: ______
Function: ______

Failure	Failure Mode		Occurrence/ Probability	Detection	Failure Effect	Severity/ Consequences	RPN class	Controls/Counter Actions		Rspnsbl	Due Date
	Component	Cause						Prevention	Detection		
Pump stopped	Pump High vibration	Failing bearing; caused by: 1. Unbalance 2. Shaft misalignment	3	2	Pumping stopped; transfer to standby	4	24	1. Assure rotor balanced at reassembly 2.Assure shafts aligned at reassembly	Implement vibration monitoring with portable equipment, quarterly interval	Dave Smith	
	Leak (controllable)	1. Gasket failed	3	2	Transfer to standby	4	24	1. Assure proper make up at reassembly	Visual inspection; hourly while in operation		
		2. Seal failing	3	2		4	24	2. Assure proper assembly			
	Coupling fractured	1. Misalignment	2	1	Pumping stopped; transfer to standby	5	10	1. Assure shaft alignment at assembly	Vibration monitoring might identify crack precursors		
		2. Defective coupling	1	1		5	5	2. Inspect coupling during assembly			
	Shaft fractured	1. Defective shaft	1	1	Pumping stopped; transfer to standby	5	5	X-ray prior to installation	Probably no warning		

Figure 12.8 Failure Modes and Effects Analysis (FMEA) template.

FMEA includes two other failure modes: partial and hidden. A partial failure is defined as a failure that results in reduced throughput or degraded efficiency. A "hidden failure" is undetected in normal operation but may seriously increase consequences resulting from a second failure. As an example, absent routine, periodic testing, failure of the start system for auxiliary or emergency equipment won't be detected until a failure of the primary. If the start system doesn't work as intended, a momentary power loss or other interruption can immediately cascade into major consequences, including a threat to life in the case of a hospital emergency power system. Likewise, the failure of an elevator door sensing mechanism won't be detected until someone sticks their hand into the gap in order to prevent the door from closing before they board. "Hidden failures" must be considered in a comprehensive risk analysis.

Used within Operational Excellence, an FMEA analysis is conducted top-down, that is, starting at the system level, progressively identifying components and equipment that could cause the system to fail.

With risk (the future) quantified and normalized for comparison with history (the past), the stage is set for the next phase within Identify—Analyze.

FINANCIAL NORMALIZATION

Finance must be involved establishing the essential normalizing factor between cost and differential production. Without normalization, many in operations/production may assert that the value of production is equal to the revenue received for a pound, ton, barrel, MW, unit, etc. As a result, increasing production and/or reducing losses nearly always appear to have large advantages over improved efficiency and cost reductions. The production first mentality. The conclusion conveniently ignores variable costs for raw materials, conversion costs (cost of goods sold, COGS), and assumes all other factors are equal such as constant pricing and undiminished demand. Here again, participation of Finance in the Operational Excellence initiative and specifically determining value prioritization is essential.

Gross margin is the business term for product revenue minus costs of production (COGS). Experience indicates operating enterprises are very reluctant to reveal comparative business details such as gross margin by production line, unit, or even facility. For the purposes of accurately evaluating improvement opportunities, similar to Figure 12.6, Finance may be willing and able to create a normalization factor that permits a valid comparison between production value and cost reduction. In terms of profit and on a normalized basis, revenue from increased production typically has value in the range between 15% and 30% of a cost reduction.

ANALYZE TO DETERMINE HIGHEST POTENTIAL VALUE OPPORTUNITIES

In addition to a Weibull or Pareto analysis to identify opportunities, financial normalization mentioned earlier is essential to value each opportunity on a common scale.

Without a normalization step, proposed improvements to processes, practices, systems, and equipment can't be compared on an equal basis.

Identify value potential, probability of success

With a normalized equivalency model in place, the next step is to construct a list of possible improvements in a sequence of potential value that will be accepted throughout the enterprise. Since Finance is typically the referee in any value disagreements between functions: mine is better than yours, it bears repeating that Finance must be involved from the outset.

Priorities are based on value to the mission/business considering the following:

- Improved SHE performance
- Improved business/mission performance—reduced variation, interruptions, lost production, inferior quality, and waste
- Improved efficiency—reduced operating and maintenance cost
- Reduced failures—reduced costs, penalties, and loss of good will
- Improved capital effectiveness—maximum output, minimum work in process, inventory.

Estimating the practicality of improvement action, degree of difficulty, cost, and time required is essential to value prioritization. Within any list of potential improvements, there will be some that are easily implemented with a high probability of success. Typically, many in this category will also be of low value. Some of the potential improvements will prove to be the most difficult and resource (primarily cost) intensive. They may also have least probability of success and take the most time to implement. In some cases, two or more improvement opportunities may produce greater value than a single higher value initiative when resources, time required, and probability of success are taken into consideration. These considerations must be factored into the prioritization process to create a refined and weighted priority list.

A facility had experienced major, costly problems in a very specific area for many years. In some cases, safety (fire and potential injury) became an issue. As the problem and possible corrective action were defined in detail, it became apparent that due to design and physical constraints, full corrective action demanded very costly modifications that would require major outage time and could themselves pose safety issues. The basic deficiency was a design flaw that was very difficult, essentially impossible to correct. Question then is what to do? Answer was to design and install an optimized surveillance system that would provide early warning of a change from normal in time to take action to prevent a well understood defect from developing into a major loss and potential safety hazard.

There is also the question of how much improvement can be gained and over what time interval. If a comparison to benchmarks reveals a significant difference between actual and benchmark performance, it is probably unrealistic to conclude that the gap can be closed in a single step change. Better to settle on an objective of a percentage improvement that participants will agree is ambitious but achievable rather than attempt to impose a major improvement that many may conclude is impossible to attain. The latter becomes a self-fulfilling prophesy. While people will take ownership and exert every effort to meet objectives they consider within reach—many won't even try to meet an objective they have concluded is impossible.

> A unit within a manufacturing facility was operating at an OEE nearly 40% below what their company had calculated as a best performance benchmark for equivalent processes and equipment of which they had many. There was no way people could be motivated to believe in a single step change increase in performance to the corporate benchmark; all considered it impossible. However, the same people would support an objective of 20% improvement above current performance.

Valuation Challenges

To emphasize an earlier statement, valuation may prove particularly difficult in the early stages of the Operational Excellence initiative due to inadequate or questionable historical data.

> A scarcity of valuation data should initiate the creation of processes to assure accurate data needed in the future is identified and recorded.

Faced with inadequate or questionable valuation data, the Program Leader should consider relaxing the valuation requirement early in the program. This might consist of a subjective opinion of value to business/mission objectives supported by justifying facts as available that can be reviewed by the Leadership and Steering Teams. Early in the program, it is far more important to familiarize participants with the process and create enthusiasm, strong ownership, and success. This can be accomplished by pursuing opportunities they identify as of high value while temporarily relaxing requirements for an accurate valuation that may not be provable with existing information. As stated earlier, missing or inaccurate information that is essential in the valuation process should initiate an improvement initiative.

FINAL STAGE

The final stage in the value prioritization step is estimating the probability of success of a proposed improvement action. In some cases, the estimated probability of success might be quite high. For others, even the most optimistic advocates may have to admit that a proposed improvement has less than 100% certainty of success. The effects: what happens if an improvement if initiative does not meet expectations, must be included in the analysis and the potential value diminished proportionately. If the probability of success is deemed less than 90%, the opportunity should be reviewed to ascertain if there are additional steps that can be taken to improve the probability of success.

Construct Priority List for Improvement

A value-prioritized list of potential improvements is constructed considering all the elements described. The list will thus represent potential value weighted by time required, cost of implementation, and probability of success.

To improve the probability of success and meeting improvement objectives, the final priority list should consist of approximately five to seven working improvement initiatives totaling approximately 120% of the results required for any given objective. Although probably best left unspoken, this will allow some slippage from objective performance and ensure the objective will be achieved even if one or more initiatives encounter unexpected delays or problems during implementation.

Conclusions reached in the Identify stage are all are preliminary estimates that will be thoroughly reviewed, refined, and potentially revised into a final prioritized listing during the Plan stage described in Chapter 14.

EXPAND LEADERSHIP TEAM

The Identify process will often provide initial guidance for augmenting the Leadership Team with additional expertise and knowledge in prime opportunity areas. Depending on the enterprise and mission, there may be requirements for experts in areas such as safety, control automation, safety instrumented systems, data management, and HR/training to name five. The awareness for increased participation should be continuous with revisions to the Leadership Team as indicated by the necessity for additional expertise dictated by specific opportunities for improvement.

As specific opportunities become clear, the Leadership Team breaks off dedicated Improvement Action Teams to pursue consensus high value opportunities. Action Teams should consist of one or more committed champions of the opportunity/opportunities from the Leadership Team augmented by equally committed individuals with essential expertise from the specific functions and areas involved. The idea is to move through Identify, Plan, and Implement as quickly as possible where permitted by improvement opportunities to rapidly gain results, demonstrating program potential and success. Progress is reported periodically to the Leadership Team.

> Correcting excessive unexpected failures is one area where the process from Identify to Implementation can be accelerated; in this case through the application of proven Condition Monitoring (CM) and Condition Based Maintenance (CBM) and Predictive Maintenance (PdM) technology. As mentioned previously, directing the application of proven optimizing practices to systems and equipment with a history of poor performance leads to greatest potential value and is one of the many benefits gained from an Operational Excellence program. Learning is accelerated by addressing the most valuable opportunities where success will be quickly recognized and greatly appreciated.

In addition to adding operating and technical knowledge, skills, and depth to accelerate and improve certainty to the process, there is another area that may require additions. The necessity to construct data acquisition, recording, and calculation processes capable of valuing improvement initiatives in terms that demonstrate compliance to enterprise business/mission objectives may also demand additions to the Leadership Team. An individual very familiar with data mining processes and dedicated software applications that might be available is almost a "must have" addition to Leadership Teams.

> The team at a site implementing a major improvement initiative found it quite difficult to identify and value prioritize deficiencies. One of the improvement team leaders developed a very innovative Excel© pivot table program capable of extracting and ranking cost data from the enterprises mainframe Computerized Maintenance Management System (CMMS). He became the site resource for data mining processes, developed a procedure that could be used by all units, and conducted training to assure all could benefit. The program was rapidly adopted by all the improvement teams, thus providing a quick, standardized method for identifying and prioritizing improvements by potential value.

WHAT YOU SHOULD TAKE AWAY

Successful Operational Excellence depends on data: data defining abnormal performance and data that will form the basis for value prioritizing performance improvements. Improvement opportunities are available in every category of the operation.

There are several ways to identify improvement opportunities, Pareto analysis of a population being one of the fastest and easiest to understand. Building the Operational Excellence program from internal data defining opportunities for improvement creates enthusiasm, ownership, and commitment for improvement. Where data-driven improvements represent the past, risk-driven improvements are an example of avoiding deficiencies that haven't yet reared their ugly head. Both must be employed.

13

PROCESS RELIABILITY TECHNIQUES HELP MAKE MORE MONEY

PAUL BARRINGER
Barringer & Associates, Inc.

This chapter uses production data of a finished product as a proxy for revenue to fund the manufacturing process and produce profit for the enterprise. Improving production processes so they consistently produce a uniformly high output makes for a predictable money machine, which functions to the advantage of investors and employees. Methods are described for how to use daily production data to find areas for eliminating problems and establish benchmarks for improvements using an understandable graphical method, with a statistical basis, for identifying and quantifying problems.

INTRODUCTION

Well-known positions exist in every factory:

1. Production "knows" most of their severe problems have maintenance roots to their issues.
2. Maintenance "knows" most of their severe problems have production roots to their issues.
3. Management "knows" from Covey's 7 Habits: "The key is not to prioritize what's on your schedule, but to schedule your priorities." Where we should work for opportunity improvements is significantly different from where we want to work.

Operational Excellence: Journey to Creating Sustainable Value, First Edition. John S. Mitchell.

Settling these adversarial issues requires quantification, in money terms, for a "To Do List" of 10 or so things that makes sense to the warring parties. Then, working on these costly items first requires good ideas for correcting the problems in conjunction with the will to solve the money issues. Keep in mind the following lost production business facts:

1. If the process *is* sold-out, then lost gross margin must be one of the cost components; however,
2. if the site *is not* sold-out, then lost gross margin is not allowed when make-up time exists.

In pricing losses from production/maintenance problems, don't make the mistake of thinking costs losses are the loss of revenue, because other costs continue during outages. Generally, production and maintenance staffs have a pretty good idea of the sales price. The gross margin is often stated as a percent of the sales prices, and it may vary from 10% to 15% for a low grade commodity product or from 50% to 75% for a proprietary product from a sole producer where percent Gross Margin = (Sales Revenue – Costs Of Goods Sold)/(Sales Revenue). This means production, maintenance, and engineering personnel need to know business costs for wise decisions, see also Chapter 6.

PARETO DISTRIBUTION

Improvement programs begin by building a Pareto distribution of problem opportunities on the basis of money—not your love affairs of what you would like to work on! Generally speaking, the Pareto distribution says 10–20% of the opportunities constitutes 60–80% of the total money issues. Clearly, it is important to work on the opportunity potentials for improvement rather than the busy work of rearranging the deck chairs on the sinking Titanic!

The Pareto distribution is named after an Italian economist, Vilfredo Pareto, who ranked the importance of problems in monetary terms in the late 1800s. Dr. Joe Juran, one of the modern founders of the quality movement, brought Pareto's concept to the factory floor with his 80-20 rule, which said 80% of the problems (that means opportunities for making improvements) will come from 20% of the causes.

Many doubt Pareto concepts because they appear too simple. Figure 13.1 shows a Pareto distribution of equipment and money from a major integrated facility. The right hand curve is based only on maintenance costs. The left hand curve is based on the sum of maintenance costs and economic losses from outages. Costs are categorized at a high level, and the curves were constructed by the Reliability Engineering manager for the purpose of persuading the organization to first concentrate on corrective action on the vital issues and to save each person's favorite love affair for resolution at a later date.

Figure 13.1 shows at point A maintenance cost for 20% of the major equipment (600 pieces of equipment) consumes 80% of the total maintenance cost. Point B is maintenance costs *plus* economic losses, which says 10% of the major equipment

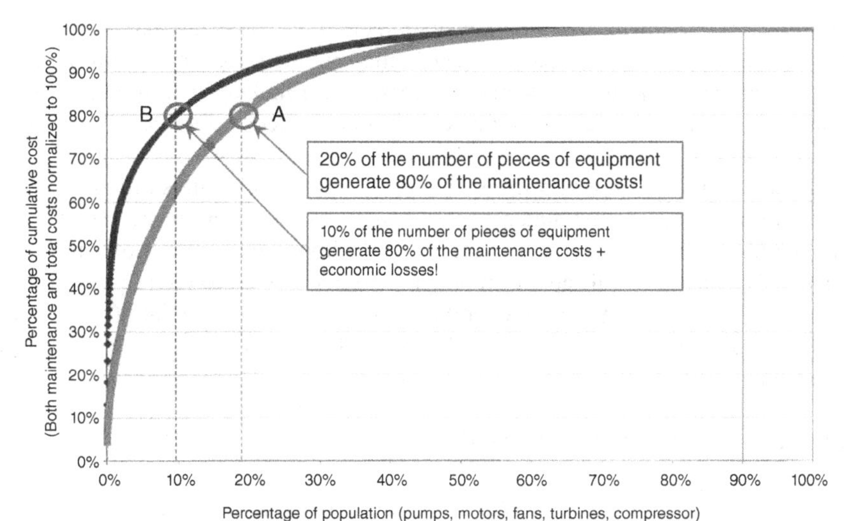

Figure 13.1 Pareto distribution curve; comparison of total costs and maintenance costs (6.25 years experience ~3000 equipment assets).

(300 pieces of equipment) consumes 80% of all costs. The 300 pieces of equipment are the vial few items requiring careful consideration and nurturing! The vital few items are different from the trivial many pieces of equipment in the upper right hand corner of Figure 13.1. Figure 13.1 identifies the few items of greatest monetary value for focusing on attacking the big monetary enemy first to control money issues—this tells us where to look for savings opportunities.

CAUSE OF DEFICIENCIES

Reasons for problems come in two broad categories as follows:

1. *special causes* are unpredictable in frequency/severity with names that you can enumerate;
2. *common causes* are statistically predictable without names attached to the problems.

Common and special causes are the two distinct origins of variation in a process, as defined in the statistical thinking and methods of Walter A. Shewhart and W. Edwards Deming. Briefly, "common causes," also called Natural patterns, are the usual, historical, quantifiable variation in a system, while "special causes" are unusual, not previously observed, nonquantifiable variation.

Generally, it's easier to solve the special cause problems because they have names and are better understood. Common cause problems do not have identifiable names and are not well understood.

Some claim maintenance and equipment problems are simpler and easier to understand, explain, and solve compared to production problems. Others claim the big money problems are in the production process, where making more products and making them more consistently would achieve a better-performing facility without the high cost issues. Both claims are correct but need quantification.

Looking at production issues from a top-down approach requires different tools than used for the top-down maintenance cost issues as shown in Figure 13.1. However, the impact will be the same by working on the important money problems first, because in business, it's all about the money. To get to the money in production, the tool to use will be a Weibull probability plot (1) using commercial PC software called *SuperSMITH*™, as the quantity of product produced each day is a precursor for money.

TRADITIONAL WEIBULL PLOTS

Engineers think about traditional Weibull plots using age-to-failure data, where the data have tails to the right or to the left and occasionally symmetrical distributions. Weibull plots developed in the 1930s transform the skewed data to a straight line for the tailed data. Engineers appreciate the use of graphical plots, where the *X–Y* plots with straight lines are easily explained and easily understood.

Weibull plots are named after a Swedish physicist, Walloddi Weibull, who spent much of his career working on fatigue failures. Weibull used his statistical graph paper for plotting test results for the Swedish bearing manufacturer SKF's bearing life test data. Bearing life data is usually not a symmetrical, bell-shaped curve, as data histograms show skewed curves with long tails to the right, as shown in the left hand side of Figure 13.2. In this illustration, the area under each probability density function (PDF) equals 1 and the *Y*-axis is a relative frequency of occurrence—notice the tails to the right. Integrating the area under the PDF curve and subtracting it from 1 produces a cumulative distribution function (CDF) curve, which has been transformed into straight lines in the right hand CDF plot of Figure 13.2. This makes it easier to understand and communicate to others—plots in Figure 13.2 were produced by SuperSMITH software.

Tailed data plotted on Weibull graph paper make a straight line with two resulting statistics: beta, β, is the line slope; and eta, η, is the characteristic life. Weibull distributions are rarely bell shaped. The eta value is not the mean, not the median, and not the mode (all typical measures related to some theme of centrality). Eta is the *X*-axis value at 63.2% probability of occurrence where the trend line crosses. It is due to a mathematical property of Weibull distributions. Knowing the eta value, all values of beta will pass through the point of eta on the *X*-axis and 63.2%, which is the horizontal dashed line in Figure 13.2 right hand plot.

PROCESS WEIBULL PLOT

Daily production output data usually makes straight lines on a Weibull probability plot. Where the steep straight line develops a cusp toward smaller production defines

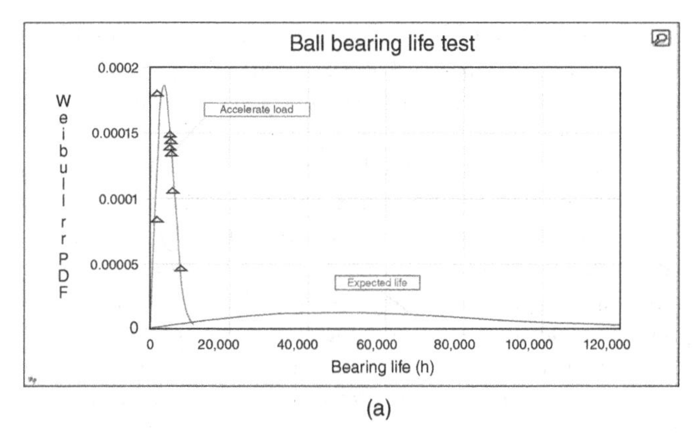

(a)

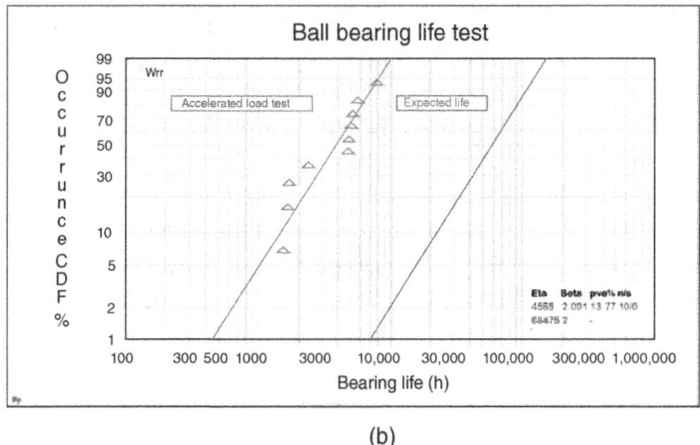

(b)

Figure 13.2 (a) Probability density function plot; (b) traditional Weibull plot.

the reliability of the process, because at that point the consistency of output is lost and bad things start to occur, such as wider variability in output along with reduced production output. The cusp defines a failure, which is always necessary for establishing reliability values. In production, we desire high reliability along with consistency in output.

Process Weibull plots have been in use since 1995, with daily output from continuous and batch processes for the X-axis. The methodology identifies variability, in a top-down view, rather than a bottom-up approach. The Weibull plot helps identify sources of problems and give clues as to why difficulties exist. It will not tell the specific "medicine" required to solve problems; however, it puts light on the issues needing correction and quantifies the losses.

Process Weibull plots help identify the amount of variability, and in many cases, it will define sources of variability. Variability comes in two categories, with *special causes* having different roots to the problem than *common causes*. Variability from special causes is easier to identify and solve. Common cause variability is more difficult to identify and solve. In general, processes with small variability have desirable large betas, whereas processes with large variability in output have undesirable small betas.

The data plots help establish the production trend lines. Loss of consistency in high output forms a cusp on the trend line toward undesirable lower output. The cusp at that point defines reliability of the process. In the field of reliability, a failure must be defined. The cusp defines the point measure of reliability at the cusp where production goes from consistently high to undesirably lower output, resulting in reduced production capacity and less money for the enterprise. Below the reliability cusp, the gaps between the production line and the actual process output can be summed to quantify production losses. These gaps of losses will be due to special causes.

The nameplate capacity of the process trend line lies to the right hand side of the production trend line, that is, greater output, than the actual production line. If you are theoretical, the nameplate line is vertical. However, production personnel will not reach for this large and frequently viewed impossible goal, thus the nameplate line must be inclined just as the thought process is for a horse to reach for a nearby carrot hung in front of the horse's nose. The production variability gap between the production line and the nameplate line represents common cause variability, problems without names for the variability. The common cause variability gaps between the nameplate line and the production line occurs for all production data points in the data set.

Waste and the Hidden Factory

Gaps of special causes and common cause sum to a wasteful hidden factory for a process; see also Chapter 15. Hidden factories represent waste with variable and fixed costs continuing for which no revenue or value is received. The hidden factory adds zero economic value because it is all waste.

Hidden factories are also opportunities for making improvements. Just as with human illnesses, no illness is cured until it is first given a name, second a root cause reason for the illness is determined, and third a remedy must be found for eliminating the illness. For a manufacturing enterprise, no waste is eliminated until it has a name and a desire to wage war on the losses and to find an effective solution.

To declare war on economic losses, there must be an objective and a strategy for waging economic war. From the objective and the strategy, tactics of the economic war are developed and implemented. Consider the few words that directly explained the Allies strategy of WWII (2):

> "The political aim was the unconditional defeat of the enemy and the return to a world of peace and order. The strategy to achieve that purpose was to take the war to the enemy by all the means that were available: aerial, land-based, naval, economic, and diplomatic. This required decisive success at the operational level, and in all the areas

covered ....they all were part of Allied grand strategy. What is incontestable, however, is that if the British, the Americans, and their smaller allies were to reconquer Europe from fascism, they first of all had to have command of the Atlantic waters."

The hidden factory war is an economic battle (not the loss of human life as occurs in most wars). As in wars of history, it requires the will to engage in battle as well as the resources to gain the advantage on the opponent. These are the same elements necessary to address and conquer the hidden factory that wastes money for the enterprise.

Special causes are economic enemies, and this enemy wears uniforms so we can identify friend and foe. Common cause enemies do not wear uniforms, and this economic enemy lurks in the open. Only by serious efforts can you identify an enemy such as Al Qaeda who wear no distinguishing labels. But they wait for opportunity to destroy just as with common causes to destroy production economies.

PRODUCTION WEIBULL ANALYSIS

World-Class Performance

Figure 13.3 illustrates the zones of special causes and common causes, which are easier to observe with poor-performing processes with many wastes. The production line defines the signal of how the process is functioning. Noise is deviation from the best-demonstrated production signal. Nate Silver (3) defines:

Signal is an indication of the underlying truth behind a statistical or predictive problem.

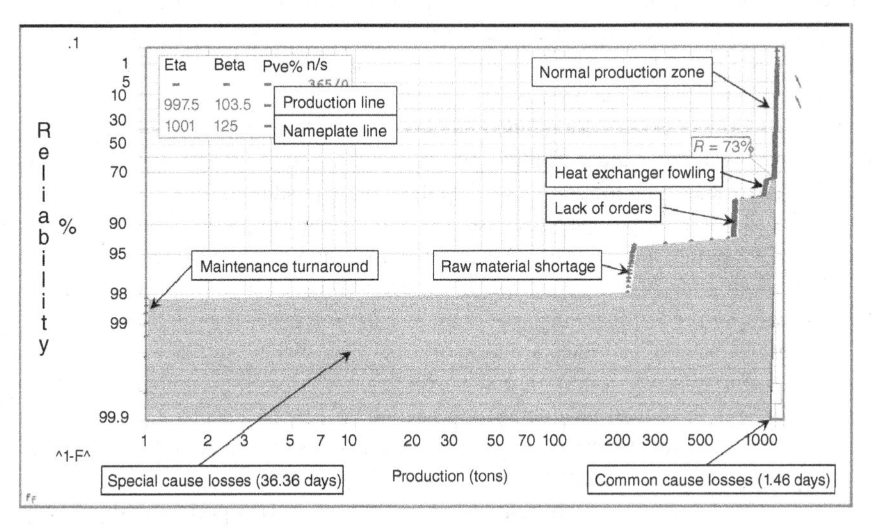

Figure 13.3 World-class production facility and process ($\beta > 100$) with losses identified in days of typical production losses.

Total production (TP) = 325,815.8 [tons/yr]
Production line (PL) = Eta 997.4661, beta 103.4687
Nameplate line (NL) = Eta 1,000.515, beta 125
Process reliability (BPR%) = 73%

Loss: Reliability loss = 36,265/997.4661 = 36.36 days of prduction lost
Reliability (RL) loss = 36,265 [tons/yr] ← Special cause losses #1 problem
73–74.58%:loss = 231.9 [tons/yr] ←Transition (0.23 days lost)
74.58–82%:loss = 2,564 [tons/yr] ←Heat exchanger fowling (2.57 days lost)
82–83.3%:loss = 1,111 [tons/yr] ←Transition (1.11 days lost)
83.3–92.7%:loss = 11,814 [tons/yr] ←Lack of orders (11.84 days lost–#1)
92.7–94%: loss = 2,474 [tons/yr] ←Transition (2.48 days lost)
94–97.9%: loss = 11.411 [tons/yr] ←Raw material shortage (11.44 days lost–#2)
97.9–98.2%: loss = 958.7 [tons/yr] ←Transition (0.96 days lost)
98.2–100%: loss = 5,700 [tons/yr] ←Maintenance turnaround (7.5 days lost–#3)
Efficiency + utilization (EUL) loss = 1,446 [tons/yr] ← Common cause losses
Total (TL) loss = 36,265 + 1,446 = 37,710 [tons/yr] = ← 37.8 days of normal production

Nameplate production (NP) = 363,533 [tons/yr]
Effectiveness = (TP/NP) = 89.62% ←Top of 1st quartile ranking of performance

Figure 13.4 Summary of special cause production losses, from Figure 13.3.

Noise means random patterns that might be easily mistaken as signals.

Notice in Figure 13.3 that 73% of all production works up and down the steep production line (the signal). Also notice with steep production lines, the exceptions show up in obvious and organized categories with clear names for the variances. All the special cause losses are equal to approximately 36 days of lost production at the typical production rate. Furthermore, for this world-class process, the gaps between the production line and the nameplate line are equivalent to approximately 1 day of efficiency and utilization losses. The clarity and definition of special causes for line slopes with small beta values (flatter line slopes) is not so clear. The old saying "The rich get richer and the poor get poorer" applies to the production Weibull's!

Figure 13.4 calculates special cause losses between the production line and the deficient data points for each step. Also notice transition zones between the different steep line segments, which occur when the process is moving from one régime to another without predominant causes for the production exception record.

The largest reliability losses are lack of orders for this 1st quartile process. This issue is an *out-of-block* condition and needs discussion with the business team, as almost 12 days of typical production are needed or 11,814 tons to fill the process. For a 1st quartile, world-class plant, marketing and sales must take orders away from lower quartile producers to fill the plant's production schedule.

The second largest loss is shortage of raw materials worth approximately 11 days of typical production or 11,411 tons, and this may also be an *out-of-block* control issue due to logistics, purchasing, or lack of on-site storage that needs to be resolved.

Lack of orders = 11,814 tons / yr * $125 / ton	$ 1,476,750/yr **Priority #1**
Raw material shortage = 11,411 tons / yr * $125 / ton	$ 1,426,375/yr **Priority #2**
Turnaround cost = 5,700 tons / yr * $125 / ton	$ 1,312,500/yr **Priority #3**
Transition costs = 4776 tons / yr * $125 / ton	$ 597,00/yr **Priority #4**
Total special cause problems	$ **4,812,625/yr** or **96.4%** of total
Efficiency and utilization losses = 1,446 tons / yr * 125 / ton	$ **180,750/yr** or **3.6%** of total
Grand total:	$ **4,993,375/yr 100.0%**

Figure 13.5 Problem-solving priority from Figure 13.3.

The last major issue worth approximately 7 days of production or 5700 tons is the maintenance turnaround, and this issue is shared by both maintenance and production for reducing lost time.

Monetizing these major losses begins with the loss of gross margin (Sales—Works Cost) for a world-class process that *should be* sold out. This 1st quartile process has 40% gross margins with sales prices of $312.4/ton, which means $125/ton will be the margin losses for every ton not produced! The out-of-pocket turnaround cost was $3,000,000 every 5 years = $600,000/year annualized. Figure 13.5 is the Pareto priority list for special cause problem resolution.

Transition costs will disappear with resolution of the major problems.

How much can you afford to spend to solve these problems? Think in terms of a simple 1-year payback and remember it's seldom that you solve all problems in each category; therefore don't overpromise and underdeliver to maintain your credibility.

4th Quartile Performance

Next, consider a 4th quartile plant of the same size and in the same business as the world-class plant. Look for the differences in performance. In this industry, first quartile plants set the sales price at no more than $312.5/ton, and frequently 4th quartile plants must give a discount of 5–10% just to get business, which would result in a sales price of approximately $295/ton. 4th quartile plants may only have a gross margin at 20%, which means the gross margin contribution is $59/ton. 4th quartile plants have many losses, and without the clarity of problem, régimes of world-class plants, output variability is much greater. Seldom are 4th quartile plants sold out, and thus they lack inclusion of loss of gross margin dollars. In short, 4th quartile plant problems are much more difficult to observe, to understand, and to remedy.

4th quartile manufacturing plants have all the same *types* of problems as a 4th quartile sports team.

1. They don't draw the best talent of coaches (managers) and team (workers) players.
2. They don't have highly paid payroll for the team (workers) and thus can't acquire the best talent (clever, innovative employees, directed to high quality production).
3. They incur many penalties (errors with scrap, waste, and lack consistency in performance).

4. Their team strategy and tactics are weak (business plan is flawed and poorly executed).
5. Team owners (stockholders) are frequently disappointed with the financial performance.
6. Team members (workers) are dispirited, fractious, and angry.
7. They never reach the playoffs (win awards and accolades for high performance).
8. They may have the same equipment and products but can't get their act together for success.

The net result of sports teams and manufacturing plants with 4th quartile performance is high variability in output, with hints of success rarely consistent in output but frequently accompanied by excuses and disappointments. 4th quartile plants lack "military discipline" of marching in precision formation, and they look very busy, but unproductive as Keystone Kops—except they are not comical as the Keystone Kops!

Figure 13.6 shows the output of a 4th quartile plant's process reliability plot using essentially the same equipment and processes as in 1st quartile plant of Figure 13.3—notice the contrast. Two régimes of special causes in Figure 13.6 are obvious (maintenance turnaround and lack of orders). Figure 13.6's label says special cause cutbacks are scattered throughout the plotted data points, which total 107 days of typical output (compare this to 36 days lost for a top performer in Figure 13.3). Also note common cause loss of 11 days in Figure 13.6 (compared 1 day lost in Figure 13.3). Setting the nameplate line for a 4th quartile plant at $\beta = 125$ as in Figure 13.3 would be viewed by the production staff as "mission impossible"

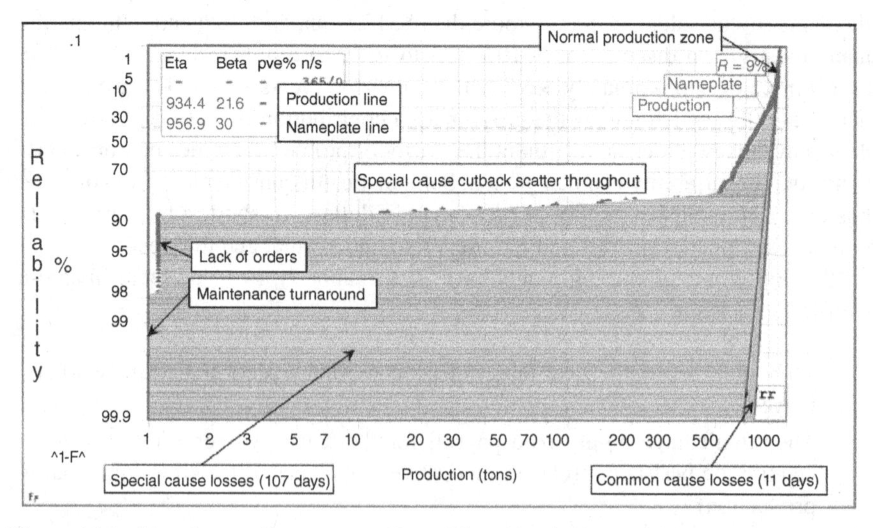

Figure 13.6 Fourth quartile process ($\beta < 25$) with highly variable output and many production losses.

for a common cause loss would be 34 days of typical production—but of course, the goal would be perceived so far beyond comprehension the production staff would not try to reach for a $\beta = 125$, as most likely they will complain about being assigned the losses of 11 days as unreasonable—but maybe achievable!!!

In Figure 13.6, the special cause reliability losses are worth 107 days of normal production. Of the 107 days lost, 68 days (68/107 = 63.6%) are due to mixed losses. How to quantify the mixed reasons by cause for dissecting specific reasons causing losses?

Weibull Probit Analysis

Weibull probit (a Weibull probability technique for SuperSMITH software) analysis (4) is the tool for quantifying the losses on the Figure 13.6 Weibull plot by the named reasons for special causes. Reasons are the predominate exception listed in the daily production report. Converting the Weibull data set of production into a Weibull probit data allows plotting of each data point, segregated by cause on the exception report, and each probit data point maintains its original Y-axis plot position for quantification of losses. Data points by each cause in Figure 13.6, and others, can be computed so the individual reasons become obvious and quantifiable.

Figure 13.7 shows mixed reasons for losses are 63338/99866 = 63.4% of special causes or roughly twice the magnitude of the next highest loss from lack of orders. Daily exception reports of explained variances in output show six major reported exceptions.

The reported exceptions in the steep section of the data plot down to the reliability point of 9% are ignored if they are reported in lower portions of the curves below the reliability point, as the reasons are only valid if never reported in the special cause categories as they belong to the common cause problems. Frequently, production operators assign spurious reasons for the lower-than-expected output in the step section of the curves, which tells they don't know the real reason for the variability—it's in the vein of "that's my excuse and I'm sticking with it."

The six major reported special causes, *in addition* to the previously identified turnaround and lack of orders for this 4th quartile plant, are as follows:

Total production (TP) = 232,806 tons/yr
Production line (PL) = Eta 934.4358, Beta = 21.50709
Nameplate line (NL) = Eta 956.9259, Beta = 30
Process reliability (BPR) = 9%

Reliability (RL) loss = 99,686 tons/yr	← **#1 Problem of 99,866/934.4 = 107 days of extra production**
9–87.5% loss = 63,338 tons/yr	← **Mixed reasons for failure = 68 days of lost production**
87.5 –98.11% loss = 31,295 tons/yr	← **Lack of orders = 33 days of lost production**
98.11–100% loss = 5,233 tons/yr	← **Turnaround = 6 days of lost production**
Efficiency and utilization loss = 10,003 tons/yr	← **Worth 11 days of extra production**
Total (TL) loss = 99,866 + 10,003 = 109,869 tons/yr	← **Worth 118 days of extra production**
Nameplate production (NP) = 342,959 tons/yr	
Effectiveness (TP/NP) = 232,806/342,959 = 67.88%	← **~ Middle of 4th quartile**

Figure 13.7 Summary of both special and common cause production losses from Figure 13.6.

1. heat exchangers,
2. instrumentation,
3. transitions from higher production to approximately 0 for lack of orders,
4. production quality problems,
5. insufficient raw materials,
6. catalyst issues, plus
7. lack of orders, and
8. turnaround.

Figure 13.8 shows the Weibull probit plot by cause. Figure 13.9 is a magnified plot of the details in the upper right hand corner of Figure 13.8. Figure 13.10 shows the summary of special cause losses by reason along with turnaround losses and lack of order losses.

Probit methodology is explained in the Barringer March 2005 Problem of the Month with different symbols for the problems as shown in Figures 13.8–13.11 of the referenced example involving product grades.

First Compared to Fourth Quartile Performance

Viewing Figure 13.9, it is clear that the untidy mixture of failures prevalent for the 4th quartile manufacturer were not evident in the data for the 1st quartile manufacturer. Figure 13.10 shows the cost details.

Figure 13.11 summarizes the cases in a side-by-side comparison, recognizing the 4th quartile plant has a basic inherent capability to be as good as the 1st quartile

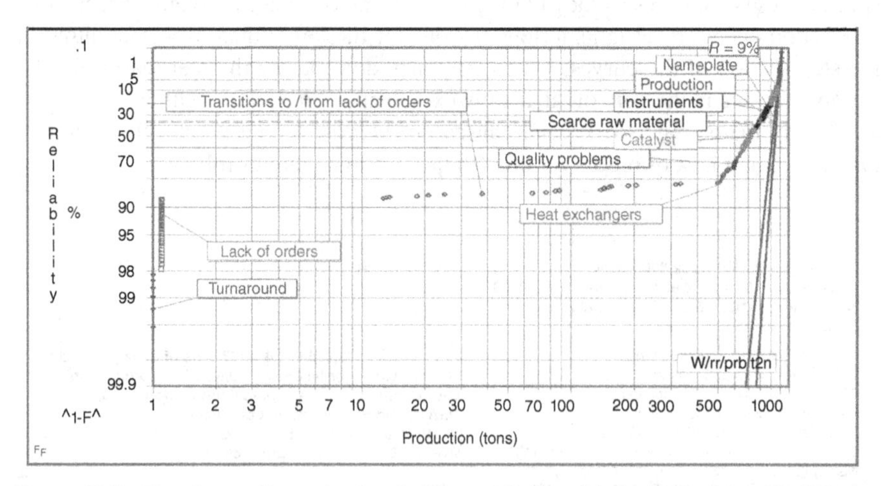

Figure 13.8 Fourth quartile production facility probit plot of daily production with different symbols for different exceptions.

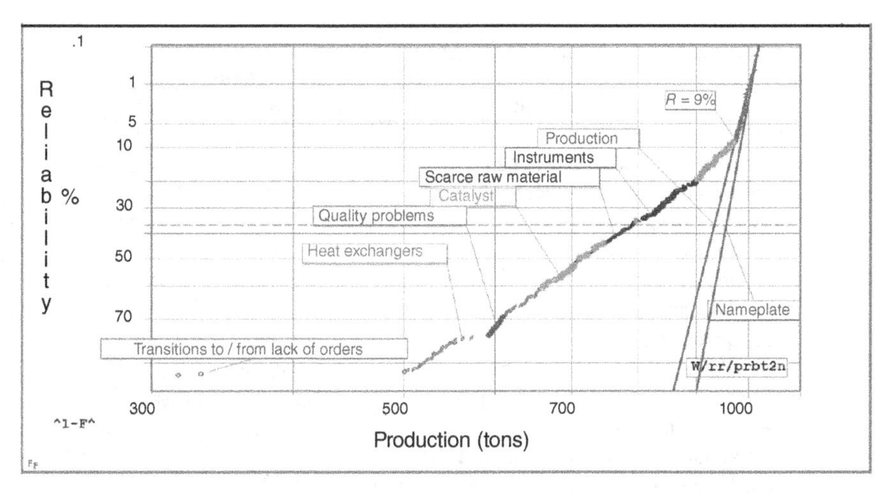

Figure 13.9 Fourth quartile production facility magnified view of mixed failure modes.

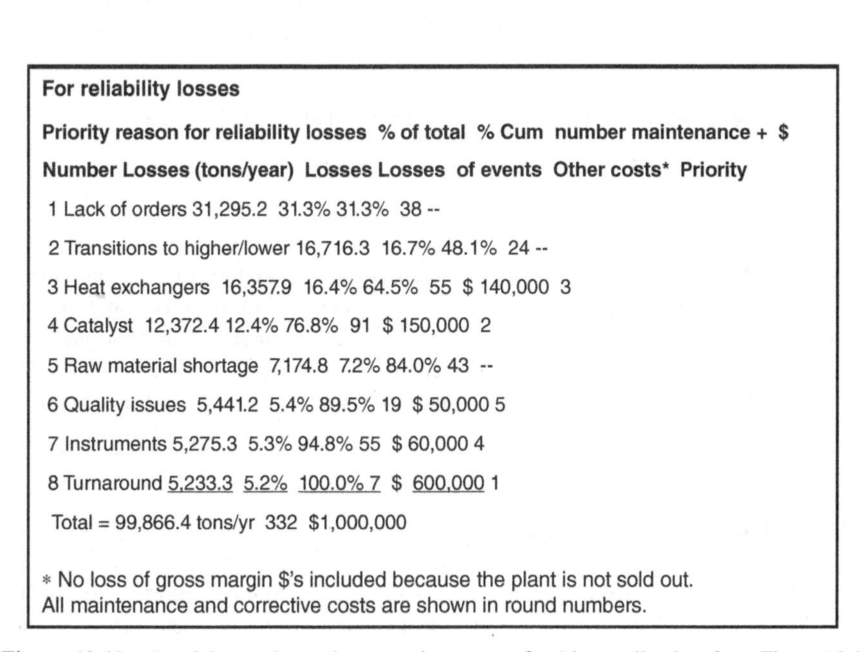

For reliability losses

Number	Priority reason for reliability losses	Losses (tons/year)	% of total Losses	% Cum Losses	number of events	maintenance + Other costs* $	Priority
1	Lack of orders	31,295.2	31.3%	31.3%	38	--	
2	Transitions to higher/lower	16,716.3	16.7%	48.1%	24	--	
3	Heat exchangers	16,357.9	16.4%	64.5%	55	$ 140,000	3
4	Catalyst	12,372.4	12.4%	76.8%	91	$ 150,000	2
5	Raw material shortage	7,174.8	7.2%	84.0%	43	--	
6	Quality issues	5,441.2	5.4%	89.5%	19	$ 50,000	5
7	Instruments	5,275.3	5.3%	94.8%	55	$ 60,000	4
8	Turnaround	5,233.3	5.2%	100.0%	7	$ 600,000	1
	Total = 99,866.4 tons/yr				332	$1,000,000	

* No loss of gross margin $'s included because the plant is not sold out.
All maintenance and corrective costs are shown in round numbers.

Figure 13.10 Special cause losses by exception reports for 4th quartile plant from Figure 13.9 in priority ranking for reliability losses.

plant when in capable hands. The 1st quartile plant has few problems to solve, but they're very worthwhile. The 4th quartile plant has many problems to resolve, but the financial advantage is small because they're rarely sold out. In the comparison,

Summary results		
	Plant	
Annual results	1st Quartile	4th Quartile
Nameplate production (tons/yr)	363533	342959
Efficiency and utilization loss (tons/yr)	−1446	−1003
Demonstrated production potential (tons/yr)	362087	341956
Reliability loss (tons/yr)	−36265	−99866
Net production (tons/yr)	325822	242090
Sales price ($/ton)	$312.40	$295.00
% gross margin	40%	20%
Cost of goods sold ($/ton)	$187.44	$236.00
Contribution to G&A + profit ($/ton)	$124.96	$59.00
Annual contribution to G&A + profit ($/yr)	$40,714,717	$14,283,310
Demonstrated process reliability	73%	9%
Potential gains for correcting all problems:		
Potential gains for correcting eff+ util losses($/yr)	$180,750	$59,177
Potential gains for correcting reliability losses ($/yr)	$4,812,625	$180,750
Potential annual contribution to G&A + profit ($/yr)	$45,708,093	$14,523,237
Potential increase to contribution %	12.3%	1.7%

Note:
These are engineering probability calculations and will not foot and tick

Figure 13.11 First quartile, 4th quartile performance, side-by-side comparison.

as illustrated in Figure 13.11, the numbers are accurate to engineering calculations but not accurate to the usual accounting values—this is due to round offs and the use of probabilities. However, the engineering numbers are accurate enough for practical problem solving in a working plant. Clearly, there is a major opportunity for solving reliability issues, particularly for a sold-out 1st quartile plant, and this requires knowing where to attack and quantifying the value in monetary terms!

WHAT YOU SHOULD TAKE AWAY

To reach maximum performance and effectiveness, it is highly important to understand losses and to have a strategy to prevent losses. It's a business problem and opportunity!

You cannot solve problems with vague generalities and good intentions. You must know where to attack and how to implement the attack on profit detractors and prioritize to solve the most important business problems first!

Time flies and losses accumulate until they are corrected. Don't wait! Build your strategy and use new tools to solve old problems!

REFERENCES

1. Abernethy RB. *The New Weibull Handbook*. 5th ed. 2006. ISBN: 0-9653062-3-2.
2. Kennedy P. *Engineers of Victory: The Problem Solvers Who Turned the Tide in the Second World War*. 2013. p 14. ISBN: 978-1-4000-6761-9.
3. Silver N. *The Signal and the Noise: Why So Many Prediction Fail—But Some Don't*. 2012. p 416. ISBN: 978-1-101-59595-4.
4. Ibid, Abernethy, pp 5-13 to 5-14.

14

PLAN OPPORTUNITIES FOR IMPROVEMENT

At the beginning of the Plan phase of the Operational Excellence program, the program and program objectives capable of contributing maximum demonstrable value to the enterprise mission/business objectives have been defined. The Leadership Team(s) has identified and preliminarily value prioritized potential organizational, procedural, and material improvements that accomplish program objectives, link to, and contribute to enterprise mission/business objectives.

As high value improvement opportunities are identified, Improvement Action Teams are formed from the Leadership Team. Action Teams have full ownership, responsibility, and accountability for validating potential value, refining, developing, and implementing detailed action plans to achieve maximum gains. Action Teams are typically led by champions from the Leadership Team. Subject Matter Experts and others with requisite knowledge, experience, commitment, and ownership are added to complete the action teams.

Although the following descriptive narrative is linear, by necessity, the actual planning process is iterative and moves back and forth. The entire process must be considered while developing action plans. Estimates of time required and cost of implementation affect final value delivered and prioritization. Both must be reviewed throughout the planning process to assure planned actions minimize time to gain results and maximize probability of success—meeting objectives. Barriers and/or considerations that potentially reduce the probability of success of a given improvement initiative may appear. Actions/activities that can be taken to maximize probability of success are added. These issues and more are addressed and resolved during formulation of detailed improvement action plans.

Operational Excellence: Journey to Creating Sustainable Value, First Edition. John S. Mitchell.

Initiatives are continually reviewed throughout the development of detailed improvement action plans. Questions include the following: what real value to enterprise and program objectives will this initiative produce? Is value demonstratively connected to and in support of business/mission objectives? What is the probability the initiative will achieve expected results and deliver expected value? As cost/time and probability of success become apparent, it may be necessary to revise the premise on which the improvement action plan is based and develop specific actions to overcome potential weaknesses and/or barriers discovered during the process.

Some of the improvement opportunities listed in the Identify phase may group naturally by similarity and applicability to a specific operating process, function, system, or equipment asset. Other linkages may become apparent, which will improve probability of success and value delivered. Similar, related, and linked improvement opportunities should be grouped for concurrent development by one or several associated Action Teams to assure greatest coordination, highest effectiveness, and maximum results.

Detailed improvement action plans for the most promising initiatives to meet program and enterprise business/mission objectives are submitted for approval, typically to the Steering Team. After approval, final plans are distributed for implementation and information.

APPOINT IMPROVEMENT ACTION TEAMS

Collaborative, multidiscipline Improvement Action Teams are formed from the Leadership Team, as illustrated at the beginning of the process flow diagram, Figure 14.1. Action Teams take ownership and prime responsibility for developing, refining, implementing, and achieving results from one or more value-prioritized improvement opportunities identified during the previous phase. Teams are typically led by individual champions from the Leadership Team with knowledge, expertise, passion, and ownership for the specific improvement(s). Teams are filled out with sufficient members to assure expertise and contribution to all aspects of an improvement opportunity. Teams should be made up of five to seven members, perhaps as many as 10 depending on the complexity of the improvement, but not many more. Size is important to promote personal working relationships between team members, necessary for greatest effectiveness. If teams are too small, there may not be sufficient knowledge or power to accomplish the objective. When teams become too large, responsibilities are too diffuse; positive, constructive working relationships may not develop. Members include Subject Matter Experts and others with knowledge, experience, enthusiasm, and willingness to take ownership for success of the initiative.

Responsibilities

Improvement Action Teams are tasked with formulating, refining, and implementing detailed improvement plans, capitalizing on specific high value opportunities. During

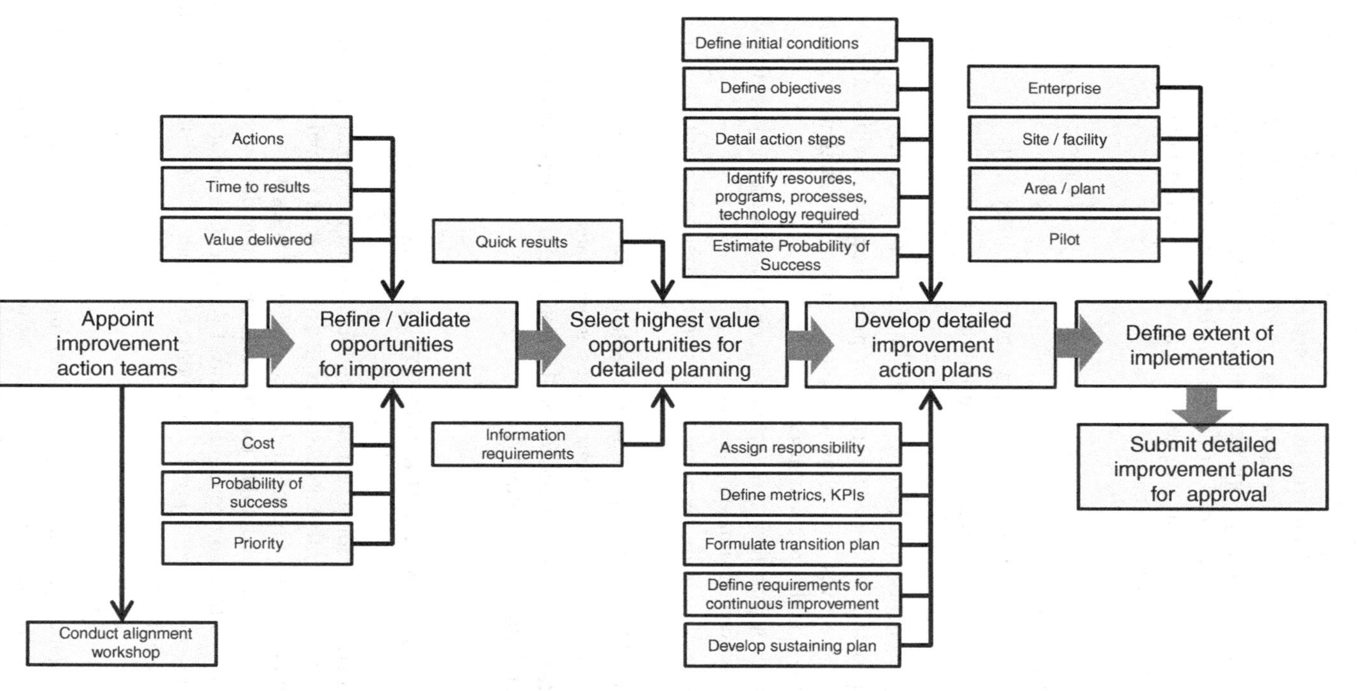

Figure 14.1 Plan stage summarized process flow diagram.

the course of their deliberations, Action Teams may identify additional opportunities for improvement that are passed back to the Leadership Team for consideration and possible addition to the identifying team or another for development.

After formation, Action Teams develop a schedule and timed milestones for refining, planning, and submitting improvement action plans for approval. Milestones for the initiative itself are a key part of the action plan. The completed action plan will be summarized in the program plan described in Chapter 9. Development in defined, sequential stages should be considered as follows:

- Refine improvement opportunity including value justification; restate opportunity if required
- Define specific objectives, future state, and preliminary KPIs
- Validate value delivered and contribution to program and business/mission objectives
- Identify strengths to build on, potential barriers, and weaknesses
- Develop detailed improvement action plan, specific activities and tasks to achieve objectives, including completion timelines, actions to overcome weaknesses, and barriers to success
- Identify resources (financial, technology, people, processes, systems, and practices) necessary for implementation
- Identify responsibility, by name for implementation and results, and create initiative RASCI
- Refine final measures of performance, activity and result KPIs with timelines
- Detail continuous improvement and sustaining plans
- Estimate probability of success; detail actions and activities to maximize probability; minimize barriers and risk.

REFINE PRELIMINARY IMPROVEMENT PLANS

As stated earlier, the preceding sequence, similar to most in Operational Excellence, is iterative in nature. It is often necessary to revisit and refine elements of the process to assure assumptions have been validated and make changes where necessary. As an example, resources required might require revisiting value delivered to assure value hasn't been eroded by resource requirements. Likewise, the initial estimated probability of success might require revisions and/or additions to the detailed action plan to increase probability.

Improvement Action Teams meet frequently, probably 2–3 days a week, during initial stages to establish momentum. Four half days might be considered if participants must balance ongoing job duties with improvement initiative activities. As initiatives gather form, meeting interval can be increased and duration reduced, as indicated by pace necessary to keep development and planning on the schedule established for submission.

During the early stages of developing improvement action plans, a representative of each Improvement Action Team should meet with the Leadership Team at least bi-weekly to assure solid communications, coordination, and knowledge of what all teams are doing and learning. Coordination is accomplished through the Leadership Team throughout the process. The Program Leader summarizes progress to the Steering Team and identifies potential challenges that might require higher level action.

Refine Improvement Opportunities

Improvement Action Teams essentially repeat the process employed earlier in the Identify phase to validate and refine opportunities for increasing value. This time, however, the analysis is much deeper, far reaching, and accomplished by the team that will have ultimate ownership and responsibility for results. If data and information aren't readily available to thoroughly define actions required for a potential improvement initiative, teams must identify and conduct detailed examinations of operating, maintenance, and other records that contain basic data from which to formulate plans and potential value gained for specific improvements, as described in Chapter 12.

The entire refinement process is dynamic, with people added to the team to meet specific requirements for knowledge and experience. Analytical methods, Weibull and Pareto analyses, have been described in Chapters 12 and 13. A value model is essential, as outlined in Chapter 5; if it doesn't exist or hasn't been started, it needs to be completed, at least preliminarily, for use in this phase.

The definition of the opportunity itself is likewise dynamic, being revised throughout the planning stage to assure actual conditions are fully accounted; the initiative itself has maximum relevancy and value delivery. Review and refinement extend to specific objectives, future state, and preliminary KPIs established during the Identify stage.

Obtain Required Information

Work during the previous, Identify phase may have led to the conclusion that valid data and information on which to evaluate and prioritize potential improvement initiatives are either inadequate, suspect, or missing altogether. In this phase, any shortcomings in data and information are exacerbated. Determining the exact cause of deficiencies such as excessive costs, failures, and inefficiency and the corresponding value of improvement demands complete, accurate, and detailed data.

Referring back to Chapter 12, the data displayed in Figures 12.4 and 12.5 were not compiled as it occurred but rather created later by tabulating information from operating logs over an 18-month period. This is tedious, certainly, but also a very valuable exercise. It revealed information to identify the source of production interruptions and hence specifics to address for improvement.

Institutional knowledge is another excellent source of information for potential improvements. Who knows more about inefficiencies and problems than the people who must contend with them on a regular basis? Action Teams must include sufficient people with operating knowledge to assure a complete, accurate assessment of conditions and opportunity. They may have to initiate detailed discussions with colleagues to expand and refine their knowledge. In any case, there is generally a wealth of knowledge defining inefficiency, defects, and potential improvements at the working level just waiting to be heard and quantified.

> During a mid-level management discussion regarding systems on which to concentrate initial improvement efforts, three were proposed as representative and critical to safe operation. When the three systems were communicated to the working level for confirmation, there was nearly total disagreement. People working directly with the systems stated two of the three were very reliable, well covered by efficient processes and procedures, and didn't require much oversight or work. They stated that two systems not mentioned, although not as critical for operation, were not reliable, had excessive and time-consuming requirements for preservation that had proven ineffective, consumed inordinate time and monetary resources, and required excessive surveillance. In this particular case, referring a management conclusion to people who really knew what was important totally invalidated 2 days of animated and often heated discussion.

Validate Value Delivered: Contribution to Enterprise Business/Mission Objectives

Although this stage in the improvement process will duplicate much of what was accomplished earlier during the Identify phase, it is essential to get the entire Action Team on board, with full understanding and total committed ownership for achieving results from the proposed improvement initiative. Perhaps more to the point, as improvement opportunities are being examined in great detail by enthusiastic, committed people who will be directly responsible for results, additional considerations, including sources of potential value add, may well be discovered.

Expanding on Chapter 12, typical sources of improving value add include the following:

- Increase Safety, Health, Environment (SHE) performance
- Increase production/mission efficiency, output, and quality
  - Improve production flow and reduce wait and dead time
    - — Improve availability, throughput, and conversion efficiency
    - — Streamline production/mission processes
    - — Reduce variation and losses: duplication, waste, scrap
    - — Minimize Work In Process (WIP)
  - Increase operating efficiency
    - — Reduce operating and utility costs
- Reduce risk
- Increase reliability—consistency of results
  - Identify and improve deficient performance
  - Eliminate defects (and the need for spending) at root cause
    - — Minimize failures and collateral damage

    — Concentrate on prevention rather than restoration
- Optimize maintenance
  - Correct design, installation, and operating deficiencies
  - Use condition-based equipment management wherever practicable
  - Minimize low value time-based equipment management
    — Review and optimize Preventive Maintenance (PM) task and interval
  - Increase work efficiency: optimize scheduled work and reduce emergency break-in work
  - Improve work precision and quality and repair success
- Improve capital effectiveness
  - Increase availability and reduce "hidden plant"
  - Reduce inventory and WIP.

Confirm Priority to Assure Greatest Return

This step also largely repeats analyses conducted during the Identify stage. Confirming value and priority assures all members of the Action Team are in full agreement with the initiative and have ownership and total commitment to success. In addition, there is the element of time. As the plan develops to attain a given improvement objective, the order in which actions can be implemented to gain quickest demonstrable results is highly important to creating quick results and enthusiasm for the program.

SELECT SET OF IMPROVEMENT OPPORTUNITIES FOR DETAILED PLAN DEVELOPMENT

Each improvement initiative should be divisible into a set of discrete actions, activities, and tasks; each of which contributes a defined, demonstrable value to the overall objective. A manageable number of highest value improvement initiatives, about five to eight in total for each overall program objective, for example, streamline processes, reduce costs, etc. are selected for refinement and detailed plan development. To repeat a point made earlier, expected value produced by the set of improvement initiatives should total approximately 125% of the overall objective (adjusted by probability of success). This adds assurance of meeting the overall objective despite less than planned results from one or more improvement initiatives.

The number of improvement initiatives that can be pursued effectively by a single Action Team depends on scope, complexity, relationship between initiatives, and difficulty of implementation. One action team can easily develop and be responsible for two to three closely connected initiatives addressing a single process or system. Too many initiatives to achieve a given objective have disadvantages in that focus, efforts, and results can be conflicting or diluted. Too few initiatives risks placing too many eggs in a single basket (a shortfall in a single initiative compromises achieving the overall result).

Prioritize Actions for Quick Results

At this initial stage, the Action Team endeavors to identify specific improvement actions that are capable of producing quick results/success to demonstrate the effectiveness and contribution of the Operational Excellence program and the specific initiative. These will be addressed first. By initially focusing on low hanging fruit—quick wins within a long-term strategy for improvement—the program gains momentum and support, both from those directly involved and from observers who see results and promise. Quick wins should have a short action horizon and produce demonstrable value and results in 3–6 months.

Directing initial focus to "quick wins" typically requires that the most difficult and visible deficiencies are addressed first. This is often contrary to instincts that tend to influence people to start small and out of the spotlight to learn a process or technology. The learning process is the same in both cases. Although the first may be more risky, it also offers far greater rewards in terms of value creation, enthusiasm, and acceptance for the program.

> There have been numerous cases where it was decided to begin a major improvement initiative in low visibility areas to minimize risk during the learning process. Since results were insignificant in terms of solving immediate problems, participants became disillusioned and found more constructive areas to spend time. Eventually, time and funding ran out on the improvement initiative without addressing the largest and most important opportunities for improvement.

Under Promise, Over Perform

Since members are selected for enthusiasm, ownership, and commitment, it is easy for Action Teams to become overconfident during the early stages of an improvement initiative. One result is to promise more than can be realistically delivered. Recognizing the necessity for early success, it is far better to under promise and over perform than the reverse.

Compensate for Inadequate or Incomplete Information

To repeat an earlier comment regarding information: in the early stages of an Operational Excellence program, the information necessary to determine an accurate valuation of one or more improvement initiatives may not be available. In the initial implementation, it is more important to create participant enthusiasm and proficiency with the Operational Excellence program rather than slowing the process to make up for shortfalls such as inadequate information. To get the program started off quickly and efficiently, accurate valuation may be relaxed for early initiatives, provided there is general agreement, including the Steering Team, that the proposed initiative is worthwhile and has tangible value and acceptable risk. Under these circumstances, it is far more important that Action Teams work on initiatives they consider of utmost importance while interest and enthusiasm are at a peak to develop proficiency with the program, maximum enthusiasm, ownership, and commitment for results.

The lack of solid information for valuing a proposed improvement initiative is in itself a learning moment. With emphasis, it leads to recognizing the importance of correcting basic shortcomings such as minimal accurate data early in the program. As initiatives develop, teams should be reminded and nudged into developing the information infrastructure that is essential to demonstrate results and value contribution to program and business/mission objectives.

> In an organization engaged in a major improvement initiative, one of the professionals suggested implementing a facility-wide optimum lubrication program. Data were not available to ascertain costs attributed to lubrication procedures currently in force, although there was general agreement the program could and should be improved and doing so would create value. The individual making the proposal became the initiative champion and was empowered to develop and implement a site-wide best practice lubrication program. Within a few months, a comprehensive training program had been completed, technicians certified, and a dedicated space procured to assure best practice receipt, storage, disbursement, disposal, and analysis of lubricants. Within a year, the program was best practice. Lubrication originated failures were dramatically reduced.

DEVELOP DETAILED IMPROVEMENT ACTION PLANS FOR HIGHEST VALUE IMPROVEMENTS

Define Starting Point

Similar to the previous stages, developing detailed action plans begins by defining the starting point: validating and refining strengths to build on and potential weaknesses and barriers to overcome. An assessment, described Chapter 17, is imperative to assure starting points are completely defined for all potential improvement initiatives.

> It is absolutely essential to define both the starting point and destination with total clarity. Think of purchasing an airline ticket: the first question is From, and the second is To!

In addition to actions directed to achieve the improvement objective, initial conditions must be considered to assure a practical transition and logical progression that provides a realistic path from current conditions to the desired future state: "where we are, to where we must be." The transition plan will be described later.

Identify Requirements for Success

An action plan contains all the information necessary to identify requirements and attain the objective results. This includes the following:

- List of applicable legal and regulatory requirements, enterprise and site SHE, risk assessment, and other procedures
- Values and behaviors essential for success and work culture requirements

- Core competencies, skills: necessity for education and training for those directly involved as well as people on the periphery who need to know what is going on as well as requirements and benefits
- Organization: detailed roles and responsibilities RASCI from facility/site manager to Improvement Action Team members
- Comprehensive communications plan to acquaint all with necessity, objective results, actions, and benefits; refer to the communications matrix in Chapter 10
- Processes and systems included in the improvement initiative
- Applicable foundation documentation, for example, P&ID, operating and repair history, and procedures
- Procedures and instructions specific to the improvement plan, including reference to applicable revision/configuration control procedures
- Training and skills requirements, including certification and re certification if applicable to demonstrate proficiency
- Transition plan: steps necessary to advance from current to expected performance and conditions including any changes required from current organization and/or procedures
- Schedule for plan review and revision if necessary and responsibility for review and updates
- Commitment to continuous improvement
- Sustaining plan to assure gains are permanent
- Scheduled assessments to confirm performance.

Detail Action Steps

A listing of all actions required without much thought given to sequence or difficulty is the most important element in the early development of a plan to gain a specific improvement objective. This will come later in the process as steps are refined and sequenced to create a path with highest probability of success. It is also necessary to identify potential "quick wins," early results mentioned previously that are highly important to create enthusiasm, motivation, ownership for success, and the all-important support for the program and improvement initiatives.

Action steps must be very specific and command oriented: do this, this, and this. Actions also must be defined in sufficient detail to assure everyone reading an action plan, including those not directly involved in development, will know exactly what is required, will be accomplished, and the expected results.

In some cases, actions must occur in a specific sequence: that is, results are required for one element of an improvement process to allow proceeding to another. This is particularly true for organizational and procedural improvements.

Each action must include a timeline, resources required, and KPIs to demonstrate progress and completion. Over the years, KPIs have been classified into leading and lagging. The former are essentially precursors to predict results that may occur significantly after the fact (lagging). In this text and as explained with examples in

Chapter 15, KPIs are classified as activity and result—essentially the same in concept but a bit more definitive. Activity KPIs are defined as those relating to processes that have to be completed to attain a stated objective and assure improvements are progressing on schedule. In themselves, activities may create little or no immediately definable value. Training is a familiar example. Training must be accomplished as part of most improvement initiatives and is absolutely essential for success. Completed training is thus an example of an activity metric. Contribution to business results, in this case results of the training, improved performance and efficiency, reduced spending, etc. are the associated results metrics. Both must have timelines for completion, producing the required results.

Define Investment and Resources

As action steps are defined, it is necessary to identify investment and resources (financial, technology, people, processes, systems, and practices) necessary for implementation. In some cases, training as an example, resources required may be primarily time. This assumes that participant's time involved in training is charged to the improvement program.

Training Training may also involve costs for third party support and possibly continuing facilitation for improvement programs such as Six Sigma, Lean and Lean Six Sigma, Root Cause Analysis, Reliability Centered Maintenance, work management, and other similar programs. In these areas, third party assistance can greatly speed up the process of implementation, assure greatest efficiency, and avoid pitfalls. All must be considered in the valuation process and divided equitably among multiple initiatives if appropriate.

Capital Equipment In addition to training, some improvement initiatives may require purchasing capital equipment and initial system setup; Condition Monitoring and Condition Based Maintenance are examples. In this specific case, equipment is typically purchased on a capital budget that affects valuation and requires a business model to demonstrate return on both the capital and continuing expense portions of the expenditure; refer to Chapter 5.

Improvements to existing processes, practices, and procedures may also require a combination of capital and ongoing expenditures. Capital for additions such as computers and computer application programs and expense for the training and facilitation to set up and implement the improvements as efficiently as possible. In some cases, it may be possible to capitalize the costs of training, facilitation, and setup.

In all cases, it is essential to accurately define exact requirements, resources required, timelines, results required, final measures of performance with activity, and result KPIs.

Estimate Probability of Success: Potential Risk and Barriers

Defining detailed actions to achieve an improvement objective in a preliminary sequence of accomplishment is the first step in constructing an action plan. The next is to estimate the probability of the initiatives success, including compliance to the timeline. What could go wrong, including a failure to meet timelines? What are the barriers and impediments to success? Actions and tasks, including regular progress reviews and KPIs, are added to address all areas affecting probability of success to gain greatest assurance that results will meet objectives on time. Participants in the process should endeavor to structure improvement plans to gain a confidence of success greater than 90%. If confidence is less, the plan is reviewed and revised with detailed actions and activities to maximize the probability of success.

Barriers of all kinds are treated in the same manner. Resistance to change is one frequently encountered barrier. The Communications Team and communications plan described in Chapter 10 are key ingredients to reduce this barrier. Communicating the necessity and benefits of improvements early and often and, equally important, successes achieved are important elements toward gaining widespread support for the improvement initiative.

With actions added as necessary to improve the probability of success and reduce barriers, the estimated probability of success is subjected to a final evaluation. Hopefully, all are in agreement that the probability of success is now above 90%.

Identify Responsibility

The plan identifies the individual or individuals responsible for implementing the plan and results. Formalizing responsibility, accountability, etc. with a RASCI chart for the improvement initiative is highly recommended.

Validate Contribution of Planned Results—Benefit/Cost—to Program and Enterprise Business/Mission Objectives

This is one of the "doubling back" steps. Do the results predicted for the improvement action plan meet the original objectives? Is there a clear sightline from action plan results to the enterprise business/mission objectives stated in the program charter? Do results comply with business/mission objective timelines? Will all interested in the initiative, particularly the Steering Team, concur with the action steps, expected results, and probability of success? All must be in full alignment. The improvement action plan must be plausible and contribute demonstrable results in total conformance to business/mission objectives.

If the value delivered isn't totally demonstrable, unquestionably contributes to business/mission objectives or appears excessively risky, participants must determine why. The initiative and action plan are revised as required to gain full compliance.

Finally, it is essential that any unintended adverse consequences that might result from an improvement initiative are identified and fully understood. Actions to avoid/mitigate any unintended consequences are incorporated into the improvement plan.

Formulate Transition Plan

When the gap between current conditions, processes, and/or procedures is large, a transition plan is developed to assure an orderly progression. A transition plan may require a series of interim steps to travel successfully from "where we are": current conditions to "where we must be": the future. Of particular importance is including everyone, so no one feels left out or vulnerable. The latter will significantly detract from the enthusiasm, commitment, and support necessary for success and must be avoided.

When multiple steps are necessary to implement a large improvement, the entire process must be mapped out in detail to assure each step builds on the last and includes necessities for future steps as well.

Develop Continuous Improvement and Sustaining Plans

With the improvement action plan essentially completed, there is one final step before submission for approval. Plans are developed for continuing improvement and assuring results are sustained. This is essentially the Control plan from Six Sigma.

Continuous Improvement The plan will contain a statement of commitment to continuing improvement on the basis of periodic performance assessments to be described later in Chapter 17. Continuing opportunity analyses and lessons learned from improvement actions are a part of the performance assessments. All involved in the Operational Excellence program must realize that standards, best practices, and competitive performance are continually improving. Eyes and attention must be kept on the future. Satisfaction is very short-lived!

Sustaining (Control) Plan A comprehensive Sustaining (Control) Plan must be an integral part of the improvement plan to assure hard won gains are stabilized, institutionalized, and do not dissipate over time. It defines demobilization and supervisory requirements after objective results have been attained. As results are achieved and objectives met, there is a natural tendency to declare victory and relax efforts. In many cases, there will be management pressure to demobilize and cease effort on an initiative that appears at the threshold of success. Pressures to relax must be resisted. Think of the Operational Excellence program as a tug-of-war; relax for a moment and all the hard earned gains can be immediately lost!

> Repeating an example quoted in Chapter 1 for emphasis: a salesman noted that he was selling an improvement program and technology into a facility for the third time! In each case, the technology was purchased to solve a specific problem or problems. Once the immediate problem had been solved, interest lagged, and eventually the program and technology were abandoned as cost reduction measures and a perceived lack of need. Within a few years of neglect, the same or a similar problem reappeared leading to the same solution. Although the salesman was happy for the repetitive sales, he asked why the facility hadn't continued with the previous, successful program. There was no institutional memory of what had gone before—success had not been sustained!

The sustaining plan begins with a sequence for demobilization as objective results are achieved. Although the action team with prime responsibility may be reduced in size, it is imperative that one or more named people retain oversight and responsibility to assure results continue to full sustainability. Full sustainability is defined as the time when there is no remaining institutional memory of any other way to perform a particular process or system. The new, optimum way is embedded in the working culture as the only way. The process, procedure, or system is no longer dependent on a champion, an individual, or individuals. At this point it is self-sustaining.

The sustaining plan has the following specific elements:

- Demobilization plan stating specific responsibilities for assuring gains achieved are maintained.
- Specific objectives: maintain improved safety and environmental performance, production uptime, reduced spending, etc. with KPIs.
- Identify any specific issues that warranted special attention during improvement implementation that must be monitored to avoid regression.
- Continuing process for periodically assessing areas of improvement. Specifics are added to the assessment procedure detailed in Chapter 17 to assure improvement results are revisited and evaluated regularly to assure continuing effectiveness and compliance.
- Identify applicable documentation that must be maintained current.
- Identify essential skills to maintain proficiency, including reinforcement, coaching, training, retraining, and certification requirements.

CONSIDER PILOT IMPLEMENTATION

As with any new concept, the introduction and first-use of the Operational Excellence program should be carefully controlled in a way that maximizes the potential for success and minimizes barriers. Operating enterprises with facilities in several geographic areas should consider a pilot program at a single facility. Facilities with multiple units at one physical site might consider a single unit pilot program.

Many of the Operational Excellence concepts may appear threatening to workers. When beginning with a pilot, fewer people are involved, and it is far easier to establish the communications and dialog necessary to gain and maintain ownership, enthusiasm, and support for a new program. Selecting a facility or unit whose personnel are enthusiastic and considered receptive to change greatly improves the chances for success. A controlled, pilot introduction thus establishes the most favorable conditions possible from the beginning to assure long-term success and the best interests of the organization.

A number of factors should be considered in the selection of facilities and units for a "pilot" installation as follows:

- Workforce receptiveness, enthusiasm, and ownership for change.
- Improvement opportunity based on mission, business, market, and operating conditions: current performance and results.

- Degree to which a facility or unit pilot program is applicable and can be transferred to other enterprise/plant businesses.
- Independence of the pilot entity providing freedom to improve processes and practices within the pilot without outside interference or veto by another because the improvements don't comply with institutional or procedural "standards."
- Reasonably close physical proximity to enterprise resources and the Steering Team to ensure continuous involvement, communications, and support throughout the pilot.

A large, multisite enterprise decided to pilot a major improvement initiative at the facility considered to be most receptive to change, although it was not the largest facility or believed to be most in need of improvement. The pilot proved highly successful and led to spontaneous appeals from other sites wanting to implement the program.

To gain a real test and achieve success, a pilot may require modifications to the corporate infrastructure, organization, and procedures. Areas that may require administrative and organizational changes to fully test pilot improvements include reporting, information flow, finance, logistics, HR, and the supply chain. Any of these and others that remain in the "old" mode can have an adverse impact on the improvements implemented within the pilot as well as diluting and diminishing results.

When it isn't possible to change the organizational infrastructure, and the pilot must operate in all or part of the old organizational environment, all potential conflicts should be identified at the beginning of the pilot initiative. With inhibiting conflicts identified, their impact on the implementation of the pilot and the results should be assessed and understood by all. The expectations and objectives of the pilot must be adjusted accordingly.

As an alternative, a pilot may be authorized to operate with modifications to enterprise administrative practices, reporting and approval procedures, or even granted permission to operate outside current administrative procedures with controls appropriate to the pilot.

Beginning with a "pilot" installation is a good way to learn and prove an Operational Excellence program. It retains flexibility, can be altered quickly to fit unanticipated conditions, is less risky, and is generally easier to sell to management. Expanding a successful "pilot" is much easier than downsizing an ambitious effort that failed to meet expectations. The latter may also discourage others from considering future improvement projects.

FINALIZE AND SUBMIT IMPROVEMENT ACTION PLANS FOR APPROVAL

Completed improvement action plans are submitted for approval, generally to the Steering Team. The full plan is presented to the Steering Team as soon as the approval process can be scheduled on the agenda. Approval presentations should be constructed for approximately 20–30 min covering highlights of which the following are most important:

- Improvement objective: contribution to program and enterprise business/mission objectives
- Final and interim objectives, time line, and KPIs
- Value delivered and other significant benefits
- Strengths to build on, weaknesses, potential risks, and barriers to success
- Summarized action plan
- Resources required–funds, technology, equipment, and people
- Probability of success, specific actions to meet potential barriers, and improve confidence of success.

The Steering Team may well ask for much greater detail. Thus, the presentation must be made by an Action Team member who possess total command of the details, is credible, good at answering questions, and not intimidated by high level managers.

Many improvement initiatives will be approved at the Steering Team level without requiring higher authority. These are typically minimal cost initiatives and/or corrective action to improve processes and procedures, eliminate problems, and inefficiencies that do not affect risk or safety.

Depending on the membership of the Steering Team, initiatives requiring cooperation between functions may require additional input for approval. Improvements requiring a new job description, additional training, and certification are examples where Human Resources involvement is a necessity.

Higher level approval will be necessary where the improvement action plan requires engineering, a capital expenditure, or affects safety and risk. If a capital expenditure is required, the plan will generally require a formal financial justification in accordance with enterprise procedures. When higher level approval is required, the initiative will be submitted by the Steering Team after their approval. Continuing engagement by the Action Team is essential.

Recognizing that improvement action plans will move at different speeds, there should be a set of completely defined and fully approved action plans totaling approximately 125% of the overall business/mission objectives ready for implementation at the completion of the Plan phase.

FOLLOWING APPROVAL

With the improvement action plan approved, it is published, communicated widely, and becomes part of the program plan. Communications, see Chapter 10 for methodology, include the following:

- A summarized description: why the initiative is necessary, and what will be done. An FAQ format may be desirable, as it can explain the initiative in short easily understood segments.

- Objectives: contribution to enterprise business/mission objectives and benefits
- Specifics of the improvement initiative:
  - Results expected: future state, value, and benefits to site/facility and employees
  - Tasks and principal actions that will be implemented to attain results
  - Metrics and KPIs: interim and final to demonstrate value gain
  - Timeline to interim and final results
  - Resources to be used: technology and training
  - Ownership, responsibility, and accountability for results
  - Other vital elements such as training requirements
  - Plan for continuous improvement and sustainability

The objective of communications is to create visibility for the improvement initiative, gain enthusiasm, support, and momentum. Communications should invite questions that are answered promptly.

WHAT YOU SHOULD TAKE AWAY

Improvement Action Teams are the heart and muscle of the Operational excellence initiative. They are appointed to address specific improvement opportunity(s), with prime responsibility and ownership to develop, refine, implement, and achieve objective results through detailed improvement action plans. Team members are selected to assure all the requisite knowledge and experience necessary to achieve success reside within the team.

Improvement action plans are fully defined with all the information and detailed steps necessary for implementation, including resources required and KPIs to demonstrate interim and final success.

During the planning process, it is quite important to estimate the probability of achieving objective results. If probability is less than 90%, the plan should be reviewed with actions added as necessary to improve probability of success.

A transition plan detailing how to move from current to objective conditions and a sustainability plan are essentials.

Finally, a pilot implementation should be considered where the step from current to objective conditions is large and/or significantly affects internal processes.

15

MEASURES OF PERFORMANCE—METRICS AND KPIs

Metrics, measures of performance, are essential contributors to the success of Operational Excellence. Metrics are objective measures of crucial program and performance parameters, including business/mission and operating effectiveness. They are the means employed within an enterprise to determine performance and effectiveness of operations, activities, systems, and equipment. Measures of performance are compared against standards to identify opportunities, create motivation for improvement, assure continuing progress to objectives, and demonstrate results.

Beginning with school grades and continuing with sports scores, league standings, blood pressure, and body weight metrics (the latter always above the benchmark!) are familiar to all. They are absolutely necessary in a modern, knowledge-based operating enterprise where survival and success demand superior performance in every activity.

Any worthwhile endeavor must have some method of assessing performance. In golf, par is a measure of performance against a course standard. By keeping score, one can determine how a player's game compares against the performance standard and other players and whether it is getting better or worse. More importantly, one can determine whether those very expensive lessons and time spent practicing are improving performance! Knowledge of body weight may determine whether that second helping of pie is a good idea!

Enthusiasm, ownership, responsibility, and commitment in an operating environment demand answers to the questions: "How are we doing; how should we be doing?" It is absolutely essential to know how you are doing on average compared to your best performance, as described in Chapter 12, as well as others with a similar

Operational Excellence: Journey to Creating Sustainable Value, First Edition. John S. Mitchell.

business/mission. Are areas of substandard performance identified; are efforts to improve and correct deficiencies and weaknesses having the desired effect?

Metrics are key links between business and financial executives and the working-level personnel largely responsible for the success of Operational Excellence. Metrics must convey the value produced. They must be layered and linked so that value from the working level cascades upward and demonstrates contribution at the enterprise business/mission effectiveness level.

Metrics can be a double-edged sword. On one hand, metrics are necessary to measure performance and establish improvement objectives; as seen in the golf example. On the other hand, incorrect or disconnected metrics can mislead and result in unexpected and suboptimal results. There are numerous expressions for the necessity of metrics as follows:

Unless you are measuring you are practicing.

You can't achieve what you don't measure.

What gets measured gets done.

People don't always use measures to get better; they may manipulate measures to look better. Bad measures can lead to poor decisions.

Without data, you are just another person with an opinion.

Within Operational Excellence, metrics must be progressive, beginning with leading, activity metrics such as training completed. This is to establish that the initial steps of improvement initiatives are proceeding as planned, with a high probability of producing the results required within the allotted time. These are followed by a combination of results metrics to measure performance as the initiative is deployed and placed in use. Finally, continued measurement and monitoring assure results are sustained and provide the basis for continuing improvement.

TYPES OF METRICS

Effective metrics are vital to assure progress to objectives and ultimate success. Metrics enable tracking, analyzing, and prioritizing activities that require the most attention, as well as areas in which improvements will yield greatest value. The sequence of measuring, analyzing, improving, and increasing objectives as each objective is achieved is the basis for continuous improvement.

A family of distinct metrics, all linked to enterprise goals, is vital to the success of any improvement program; certainly Operational Excellence. Metrics are necessary to identify opportunities, prioritize resources, and measure results—progress and effectiveness of improvement initiatives. Metrics help to better understand the contributors to enterprise business/mission success, as well as the impact of availability, throughput, quality, and cost.

Metrics can be classified in several ways. For the purposes of this discussion, metrics are classified into two groups: activity and results. Activity metrics must represent improvements that will lead to results and value. It is very possible to have a great deal of activity without producing results. Observe 6-year-old children at school recess for an example!

Results metrics must focus on value added. Value added by improving performance to objective levels.

Activity Metrics

Activity metrics typically measure progress to completion and are applied to actions taken to improve skills, proficiency, processes, organization, etc. Hours studying and training; introducing technology and improved processes, and reorganizing and optimizing flow to increase efficiency and progress to improvement milestones are examples. Golf lessons are a typical example of an activity designed to improve skill and proficiency. Activities in themselves may not produce value; most always, they are essential prerequisites for creating value.

The objective of most activities is completion. Using deployment of a new or improved process, Operational Excellence as an example—the initial activity metric might be percentage of training completed. As training is completed and Operational Excellence deployed, the activity metric would shift to program objectives and organization established, multifunction improvement teams formed, and finally improvement initiatives identified and developed. Each of these categories would have subsets as described in preceding chapters, all measured in percentage completion. Note the activity measurement/metric would not include improved results. As the name suggests, this would be a results metric.

Results Metrics

For the most part, rewards aren't given for activities or trying hard but are given for results! Results metrics measure performance and contribution to an objective—value in the case of Operational Excellence. Reusing the golf example, score is the results metric. A continuously declining score indicates improvement, greater skills, and proficiency. Whether it justifies the cost of lessons is up to the golfer to decide. School grades, athletic scores and standings, weight loss, and speed records of all kinds are other familiar examples of results metrics. Improving grades, running faster, declining weight are the results, the pay off of studying, training, eating less, and exercising; activities all.

Results metrics can be further subdivided into two groups. Many results metrics are generally applicable across industry groups. Most maintenance work management metrics are in this category. Others, such as cost per unit production and MTBF (mean time between failures), are industry and even equipment specific. The differentiation is somewhat arbitrary, and the metrics tend to overlap. Linkage, flow, and consistency of purpose are of prime importance.

Results metrics are available from many sources. International and national standards, professional organizations, and published literature are primary sources for terminology, definitions, and benchmark values.

Results metrics typically require a numerical objective. A numerical objective can be performance to a published benchmark; absolute or percentage improvement from current performance.

Results metrics within Operational Excellence are generally improved performance exemplified by improved SHE (safety, health, and environmental) performance, business/mission effectiveness, increased availability, reduced costs, unscheduled downtime, etc. Reusing the golf example, performance to par, or subpar in the case of a professional, would be a typical results objective, par being an example of a published performance benchmark.

Many at the working level question whether universal benchmark values can or should apply across an industry in general or even similar industries. An often-stated excuse for poor performance is the benchmark objective was derived for a different facility or process.

This leads to using a percentage or numerical improvement from current performance as the objective. Examples include improve availability for production/mission by 20%; reduce costs and/or unexpected failures by a stated percentage. There are many benefits. Objectives of this type are immediately understood and credible to all who will be working to achieve the objective. They often lead directly to ideas for improvement and will generate the essential enthusiasm, ownership, commitment, and responsibility for success.

Using a percentage improvement from current performance removes the argument that the benchmark doesn't apply equally to this facility or process. It is equally important that percentage improvements do not compromise numerical information that the enterprise may consider proprietary.

Leading and Lagging Metrics

Many mention leading and lagging metrics. Leading metrics are typically activities undertaken to produce a given result. Since the result generally follows the activity, the result is generally referred to as a *lagging metric*. For the purposes of the Operational Excellence program, metrics will be called what they really are: activity and results.

ESTABLISHING OBJECTIVES

It is crucial to establish realistic objectives. Improvement Action Teams should be encouraged to establish optimistic "stretch" objectives to gain the highest realistic performance and results; perhaps a bit more than people are comfortable with. Recognize, however, if objectives are set too high, then they may be considered unrealistic and improbable, causing many to give up and not even try. Thus, establishing objectives whether as an improvement from current performance or considering a performance benchmark calls for a great deal of thought and careful consideration. As has been stated earlier, it is better to underpromise and overperform than to overpromise and underperform!

Inclusion and participation are essential to develop the optimism, ownership, and commitment necessary to achieve success. Allowing working-level improvement action teams—people with actual responsibility for results—to establish performance objectives typically leads to more ambitious objectives than anything that could be ordered from above. It also promotes strong ownership and a high level of commitment to attaining objectives.

Benchmarks

Industry best practice benchmarks are a source of objectives for results metrics. There are many industry best practice benchmarks published and widely available. These include SHE performance, capital and cost-based measures of business/mission effectiveness, customer response, production effectiveness, quality, efficiency, logistics management, maintenance work, and other process-specific benchmarks. All provide a means of comparing actual performance to industry best.

Benchmarks are typically classified into four categories: industry, program, process, and function. Each industry has generally accepted overall "world class" benchmarks that are useful for determining comparative performance, for example, cost and value of spares inventory as a percentage of RAV (replacement asset value)/ERV (equivalent replacement value) (explained later in this chapter) or equivalent distillation capacity (EDC) in the petroleum refining industry, tons of steel production per availability hour, and assembly hours per automobile. From these, internal enterprise or facility metrics can be established for the Operational Excellence program for prioritization and to drive improvement initiatives. Many program, process, and functional benchmarks apply across industry groups. As stated earlier, most maintenance work management benchmarks are in this category.

Mission, intensity of operations, location/environment, and age must be considered for an accurate comparison of performance to benchmarks between enterprises and sites. Plant age is a factor. After infancy problems are corrected, a new facility or process will typically be more efficient, perform better, and incur lower costs than an older equivalent. A facility, process, system, or equipment asset operated close to or above nameplate capacity will generally cost more compared to equipment, a process, or facility operated less intensively. Facilities, processes, and equipment operated in hostile environments will generally be costlier and more prone to failure compared to those operating in a benign environment. Finally, geographic areas can have a major effect. Although some costs may be higher in facilities located in the Middle East and Southeast Asia, labor costs are typically much lower compared to other areas of the world. In each case, differences must be considered when establishing performance objectives for Operational Excellence.

Benchmark values can be used for guidance and comparison; however, for the reasons cited previously, they are not always absolutes that can be applied with confidence across industries and processes. Care must be taken to ensure that objective values established from benchmarks are considered realistic and attainable by all involved and responsible for compliance. Whether derived from benchmarks or other sources, all who will be involved in an Operational Excellence improvement process should participate in establishing precise objectives, methods for achieving objective results, the corresponding metrics, and methods of calculation. Objectives that appear unrealistically high and out of reach will not gain the ownership and commitment necessary for success.

USE OF METRICS

Metrics are essential to identify improvement opportunities and convey the corporate value and return gained by improved processes and technology through Operational Excellence. Defined and structured properly, they provide the connection from improved performance and effectiveness at the working level, to contribution to the enterprise overall business/mission strategy and objectives.

Program metrics are used to:

- Establish and monitor performance and effectiveness—compare performance to benchmarks and demonstrate contribution to business/mission objectives.
- Identify opportunities for improvement—means to communicate necessity for improvement.
- Strengthen best practices—identify and eliminate poor practices.
- Compare/monitor performance as improvement initiatives are implemented to assure success—is performance getting better?
- Form the basis for justifying virtually all activities within Operational Excellence, including the sustaining investment of personnel, resources, technology, and training required for continuing, permanent success.

Metrics must be selected to accurately represent elements of performance that clearly contribute and link to overall business/mission results. They must be relatively simple and easily understood—school grades, golf par, and body weight examples cited earlier. Measurement and calculation method must be objective, quantifiable, repeatable, and unambiguous. Data collection, calculations, and distribution must be automated for consistency and to minimize time and the administrative burden. Metrics must be measured at regular intervals, generally monthly for most metrics. Metrics are reviewed and published periodically; revised and improved as necessary.

There are several basic considerations to be addressed when establishing performance metrics, which are as follows:

- What is the metric intended to prove/demonstrate?
- What is its definition; are there accepted business/industry/process equivalents?
- What is to be measured; what is the calculation process; can it be automated?
- Where are the data; are they readily available and measured consistently and accurately?
- Who is responsible for the measurement(s)?

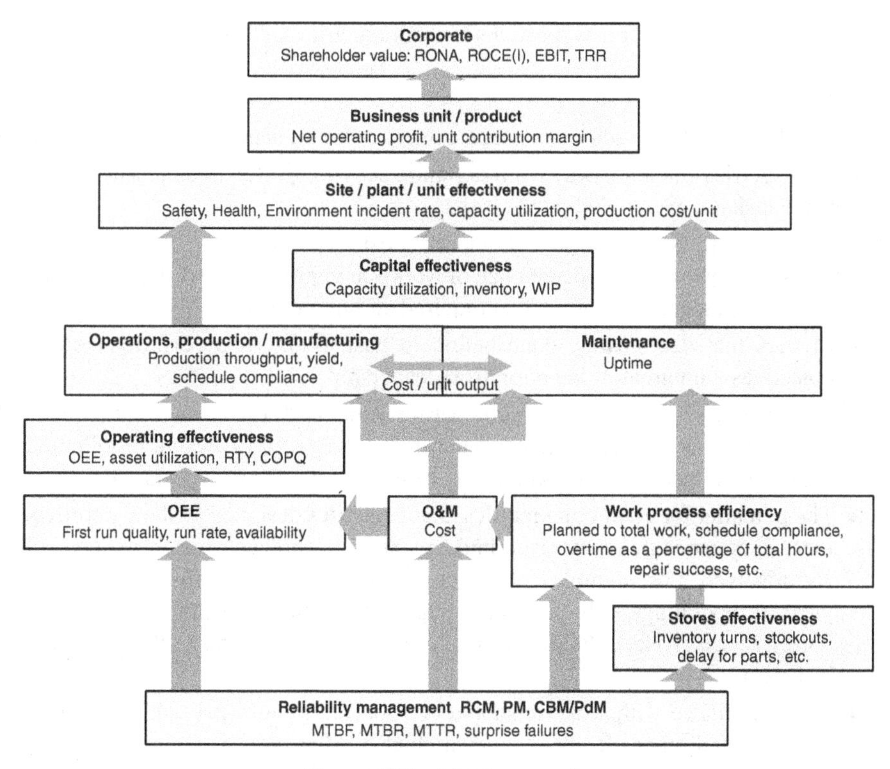

Figure 15.1 Metrics cascade.

- How and to whom will the metric be reported?
- How will it be used?

The Operational Excellence strategy is driven by the enterprise business/mission objectives and strategy. This requires overall and performance effectiveness metrics that are linked and cascade from bottom to top and top to bottom, as shown in Figure 15.1. The cascade assures that program objectives are consistent and linked. Most important, all results of improved performance and effectiveness gained through Operational Excellence connect to and convincingly demonstrate value toward business/mission objectives.

Note that in Figure 15.1, each set of metrics links to and supports the layer above.

From this brief description, it is easily seen that multiple metrics are required to assess and communicate the performance, effectiveness, and contribution of a complex, comprehensive Operational Excellence program.

Requirements

Metrics must be concise and limited in number. Focusing on too many areas at once may result in lack of emphasis and information overload. Equally important, it may diminish the ability to direct limited resources to highest value activities.

The following are several suggestions to ensure metrics are developed and applied effectively:

- Good metrics focus activities to achieving maximum benefits and value added.
- Poor metrics can lead away from optimum activities, often to emphasis on low value tasks or even unintended results.
- Whenever possible, metrics should be positive, rather than negative (e.g., measure work quality—percentage of work accomplished correctly, rather than rework—percentage of work that required additional work).
- Metrics that demonstrate contribution to business/mission performance and objectives gain greatest attention from leadership.
- Metrics must be realistic when comparing actual performance against requirements. In many cases, requirements may have changed since design/installation. A presumed lack of performance may be due to unrealistic expectations.
- The potential for conflicting metrics, that is metrics designed to show improvement in one area that may result in degradation in another, must be understood and resolved. For example, there's no benefit in directing efforts to increase yield if operating at higher rates degrades quality to significantly below objective. Likewise, reducing inventory may also reduce turns of the remaining inventory.
- Noncompliance with a metric should be followed by efforts to identify cause, full cost, and other effects of noncompliance. Many organizations use Pareto analyses for this purpose; refer to Figure 12.4.
- Metrics must be used and kept current. Metrics that are not used regularly should be eliminated.
- Sustaining metrics must be established to assure performance improvements do not decrease due to a lack of attention.

Effective metrics measure results that those responsible can control and improve. Span of control, ownership, and contribution are vital to the overall improvement and optimization process.

> Metrics develop awareness of the need for improvement. Without accurate comparative data, entities may think they are doing well when the opposite is closer to the truth.

Metrics serve as the benchmarks from which to identify opportunities for improvement. Objective performance can be published industry benchmarks or a percentage improvement to current performance, as suggested in Chapters 12 and 13. SHE performance is typically numerical; hours since last lost time or recordable accident as examples. Production/mission effectiveness is usually expressed as a percentage. Availability can be expressed as a percentage or in time. Percentage may make more intuitive sense; 95% annual availability is more meaningful than

8322 hours. At the working levels, measures of performance are often measured in percentages—planned as a percentage of total work is one example. Schedule compliance and work quality are others.

In every case, the first step is to identify the five or so most important metrics at each level. Figure 15.1 includes suggestions. Accuracy and repeatability demand clear definitions and a consistent data acquisition and calculation procedure.

With definition and calculation procedure established, the data required must be located within the enterprise information system and the calculation procedure established. If the metric is for comparison with an international or industry standard and benchmark, every effort must be made to assure the definition, data, and calculation procedure are identical to the process used by the standard. Most important, once a definition and calculation procedure have been established, it is absolutely essential to maintain consistency year after year.

Metrics must be seen as a positive force driving improvements that are beneficial to all and not attach blame. This is an essential step that assures progress. Otherwise, metrics will be manipulated to ensure that performance, accurate or not, is as high as possible.

An operating enterprise was mystified by continually improving availability of one business unit, while other measures of effectiveness were declining. Further study determined that the availability objective for a year was established on the basis of the previous year being 100%! In this instance, 96% each year actually represented a 4% year-to-year decline!

When compensation is involved, be assured that every effort will be made to alter numbers wherever possible to paint the brightest picture possible!

All key processes and practices must have at least one result and time metric to identify the objective. Likewise, each metric must have an owner who is responsible for meeting the objective(s).

If the data necessary to calculate essential performance metrics are not available at the commencement of an Operational Excellence program, they must be developed and implemented before improvement initiatives are begun. This will ensure that improvement initiatives will build the history that is essential to conclusively demonstrate value.

A facility 2 years into an improvement initiative had not implemented any data-gathering process to demonstrate the effectiveness and value created by improvements. As a result, they could not demonstrate effectiveness and were concerned that 2 years of valuable history proving the success of improvements had been lost.

Characteristics

Metrics must be specific (unambiguous), measurable, understandable, directed to enterprise business/mission objectives, actionable, and controllable by those charged with compliance. At the business and financial executive levels of an enterprise, capital and profit-based metrics: RONA (Return On Net Assets), ROCE (Return On Capital Employed)/ROCI (Return On Capital Invested), EBIT (Earnings Before Interest and Taxes), and their derivatives along with share price are of greatest importance. Business/financial level performance and effectiveness metrics will be established by enterprise executives.

At the program level, metrics are used to measure the performance and effectiveness of specific processes, systems, practices, and equipment, identify deficiencies, and provide the value basis for improvement. Every program objective must have one or more well-defined activity and results metrics, typically multiple metrics with time lines and actions required to keep everyone engaged and focused to gain compliance. Used largely by the Steering Team and Improvement and Action Team leaders, these metrics identify opportunities and measure the performance of the ongoing optimization and continuous improvement process.

At the program and working levels, business and financial metrics such as RONA have little meaning and seemingly aren't controllable. Thus, it is imperative that Operational Excellence program metrics connect directly into and demonstrate value at the business financial level. Linked metrics provide a path demonstrating how characteristics that can be controlled at the working level—increased availability and reduced failures to name two, connect to and improve business/financial results. Refer again to Figure 15.1.

Metrics must be easily comprehensible. All with responsibility for controlling performance must know exactly what composes the metric and how it is moved.

In several facilities, workers were able to quote the current value of a safety metric yet had little or no understanding of what the number represented.

Definitions

A valid comparison of performance measurements requires consistent definitions and rules. What is a failure? How is a partial failure defined if the system or asset continues in operation in a degraded condition? What is downtime? Is a facility, system, or equipment asset that is in standby considered down? The causes of downtime—market effects (lack of sales/demand), operational considerations, process difficulties, and equipment malfunctions must be fully identified. Without properly identifying cause, availability and reliability values will be inconsistent and may lead to incorrect conclusions and erroneous prioritization of improvement efforts. ISO14224 is a good starting point for guidance and answers to all these questions. It may not be entirely applicable to specific circumstances within a given operating facility, but it will form an excellent basis for establishing an optimum structure and definition. See also Figure 12.3.

During a detailed discussion of metrics, one participant commented: "how can anyone expect agreement on performance metrics when we don't have an accepted definition of failure?" A suggestion was made that any condition requiring system or equipment lock out would be considered a failure.

Whenever applicable, definitions must conform to accepted standards and conventions. This includes the metric itself, data basis, and the method of calculation. The definition and method of calculating business and financial metrics, RONA, ROCE, etc., are fully defined by national and international accounting standards. Many process effectiveness metrics such as OEE (Overall Equipment Effectiveness) and those defined for maintenance and work management are similarly defined by professional organizations. All who will use a given metric, especially those accountable for results, must agree to consistently use a generally accepted definition and calculation methodology whenever applicable. If a generally accepted definition or method of calculation doesn't exist, it must be proposed and agreed on by all who will be affected. Definitions should be totally documented and readily available to ensure consistency of application.

> After many attempts to define failure, one operating facility defined an event as any anomaly that required the system or equipment to be locked out. They made some exceptions, changing belts as one example. Essentially, the change in terminology from failure to event resolved the controversy.

Establish Magnitude of Value/Opportunity

Whether the goal is a percentage improvement or improvement to a benchmark, the difference between current and objective performance establishes the potential value gain for closing the gap. Ultimately, this will be used within Operational Excellence to prioritize improvement initiatives for implementation. Thus, this step—based on metrics and described more fully in Chapter 14, is crucial to the process and necessary to direct attention to highest potential value opportunities for improvement.

Communicate the Need for Improvement Metrics provide an easily understandable method of communicating the need for improvement—both visually and in words. Metrics are the score; in virtually every endeavor and every competitive activity, the score is all important, known to all, and incentive to improve. Everyone should agree that SHE incidents, quality defects, failures, and forced outages are bad in any operating enterprise. Displaying a graph of forced outages over time in terms of outage time or events immediately communicates what all will conclude is a problem, as well as justifying the necessity for corrective action and demonstrating improvements.

Identify Strengths to Build on and Weaknesses to Correct The comparison of actual performance to industry and process benchmarks will identify both strengths and weaknesses within the enterprise. Performance close to equivalent industry best indicates strengths to build on. Performance well away from industry best identifies weaknesses that must be corrected and in some cases demands a step change improvement. In each case, metrics and a consistent calculation method form the basis of comparison.

Display Performance, Progress, and Results to Objectives Graphs, visual communications, are an excellent method of communicating progress to objectives

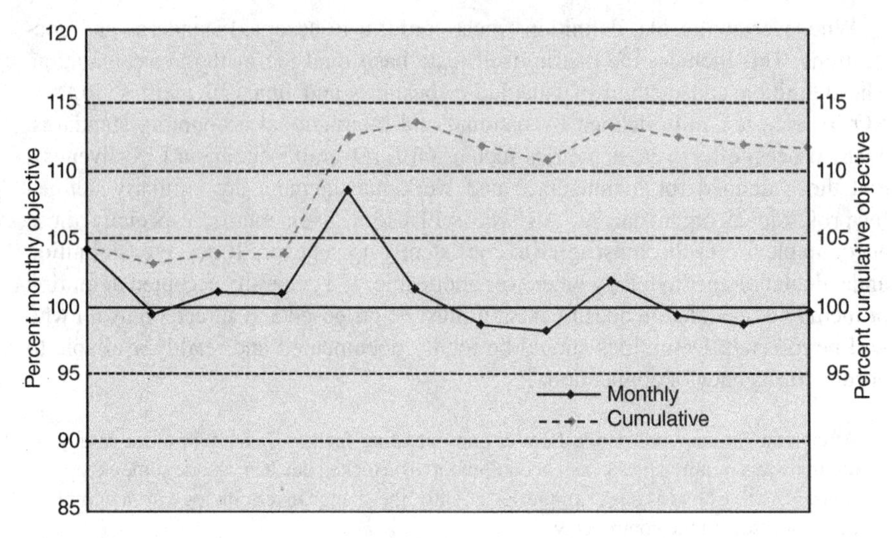

Figure 15.2 Monthly and 12-month cumulative performance plotted along with the performance objective.

throughout the enterprise; refer to Figure 15.2. Actual, both monthly and 12-month cumulative performance, plotted along with a benchmark or performance objective, 100% in this case, provides a powerful display of the improvement opportunity—the gap that must be closed.

Confirm Progress Toward Slow Moving Overall Objectives When improvements are made to individual processes, systems, and/or equipment within a large population, results, changes in an overall metric, may not be readily apparent. This can be rectified by creating and monitoring a specific metric for the segment of the population on which improvements have been implemented. Comparative performance before and after improvement is more readily observable.

Develop People–Raise Awareness, Ownership, and Commitment As the score of the success of an Operational Excellence program, metrics raise awareness, ownership, and commitment. Everyone understands scoring—whether we are ahead, behind, or simply holding our own. Referring back to Figure 15.2, it is immediately apparent that a little over 10% improvement is needed to regain objective performance. Metrics properly configured, communicated, and displayed keep progress to objectives front and center.

Assure Continuous Improvement Identifying opportunities, metrics and gaps to objective performance form the basis for continuous improvement, as well as demonstrating progress to improvement. As an improvement initiative nears completion, continued tracking of key metrics demonstrates the improvement is being sustained.

HIERARCHY OF OPERATIONAL EXCELLENCE METRICS

In general terms, all performance metrics must be layered and linked in a hierarchy to demonstrate value to the metric directly above. Referring again to Figure 15.1, it must be possible to drill down from a dashboard of top tier metrics through a hierarchical stack with increasing explanatory detail at each level. This enables identifying location and cause of deviations, as well as crediting improvements. For example, a low value of RONA caused by inefficient production (availability, run-rate, and quality) or excessive cost. If low availability is the culprit, where is the cause and what is it? Likewise, cost—where are the deviations and what is the cause(s)? In both cases, the metric hierarchy must be structured that will allow identifying specific processes, systems, and equipment responsible for deviations and the specific cause of the deviation.

With this emphasis, employees at all levels in the organization will understand the metrics for which they are responsible and why the metrics are important.

In an operating enterprise, the hierarchy begins with SHE performance metrics. To reemphasize a point made earlier, an enterprise cannot achieve Operational Excellence or excellence in any area without achieving SHE excellence.

In an operating business, SHE excellence is immediately followed by a top-level measure of business/financial performance and effectiveness such as EBIT, RONA, ROCE, ROCI, or a variation. Reviewing from Chapter 5, earnings and return on capital demonstrate the corporation's ability to create value. They are probably the only rational link between business performance and shareholder value. It is thus essential for this level to cascade and link into the production, process, and system effectiveness metrics that create earnings and return on capital. Business value and effectiveness metrics consider the market and competitive position, as well as operating effectiveness. A declining market and/or competitors applying price pressures will decrease return—the numerator in each case.

Operating effectiveness metrics followed by program effectiveness metrics are next; selected to demonstrate contribution upward in the hierarchy from actions at working levels. Working-level effectiveness metrics are selected to measure the contribution to profit and value (production/mission capability), identify the origin of deficiencies and costs (as the basis for improvement initiatives), and verify the effectiveness and return gained by improvement initiatives.

A mission-oriented operating enterprise must have top-level mission and strategic compliance metrics linked to operating and working-level metrics similar to those described for a manufacturing/production enterprise.

Operating Effectiveness

With the exception of SHE, operating effectiveness is the first level in the hierarchy that is generally under control of the operating unit. Other metrics controlled by the operating unit include the following: cost per unit produced, OEE, OOE (Overall Operating Effectiveness) from Chapter 5, capacity utilization, lost opportunity,

Rolled Throughput Yield (RTY) and Cost Of Poor Quality (COPQ) both from Six Sigma, and uptime (ISO14224).

In terms of Operational Excellence, measuring cost effectiveness in terms of cost per unit output rather than output and cost separately is the best metric that points efforts in the right direction. In addition to the obvious of seeking improvements in effectiveness by reducing cost per unit, increasing costs can also be justified to produce more if the increase in production reduces cost per unit or at least keeps it unchanged. This assumes a stable market that will accept the increased production without appreciable price erosion.

A manufacturing enterprise in a market where everything produced could be sold without any price erosion established increasing production while holding fixed expenses constant or nearly constant as the primary strategic objective.

Cost as a percentage of RAV or ERV is a frequently used metric for representing cost effectiveness in the downstream hydrocarbon industries. It was originally proposed as a broadly applicable, neutral measure that could be used as a normalizing factor across plants and processes. By the early 2000s, cost as a percentage of RAV/ERV had gained visibility at the senior management level and was being applied across a broad segment of industry as a top-level benchmark for maintenance effectiveness.

RAV/ERV is defined as the current cost to replace production capacity and is considerably different from depreciated book value. The challenge is that the cost to replace production capacity is quite difficult to determine accurately, as processes have changed and production facilities have typically increased in size and efficiency (increased production output at less capital cost). All typically resulting in reduced costs per unit produced.

If cost as a percentage of RAV is to be used as the measure of effectiveness, the best method appears to be to define RAV at one point in time and then use the value for the denominator over an extended period with an objective of reducing the percentage. If changes are deemed necessary to the calculated RAV, it should be applied to last data along with the previous value of RAV so the offset can be understood. A bit analogous to changing to a bathroom scale that weighs lighter; the changeover didn't cause any real loss in weight!

Overall Equipment Effectiveness OEE is a commonly used metric that originated from the Toyota manufacturing system. OEE is the product of availability (uptime), production throughput, and first-run-quality (yield). Each term is expressed as percentage effectiveness of actual divided by objective performance. World-class performance in terms of OEE varies from approximately 85% for discrete and batch manufacturing to considerably higher for continuous processing operations.

OEE must be used with one caveat: availability must coincide with operating requirements. This is especially important when requirements to operate are less than 100%.

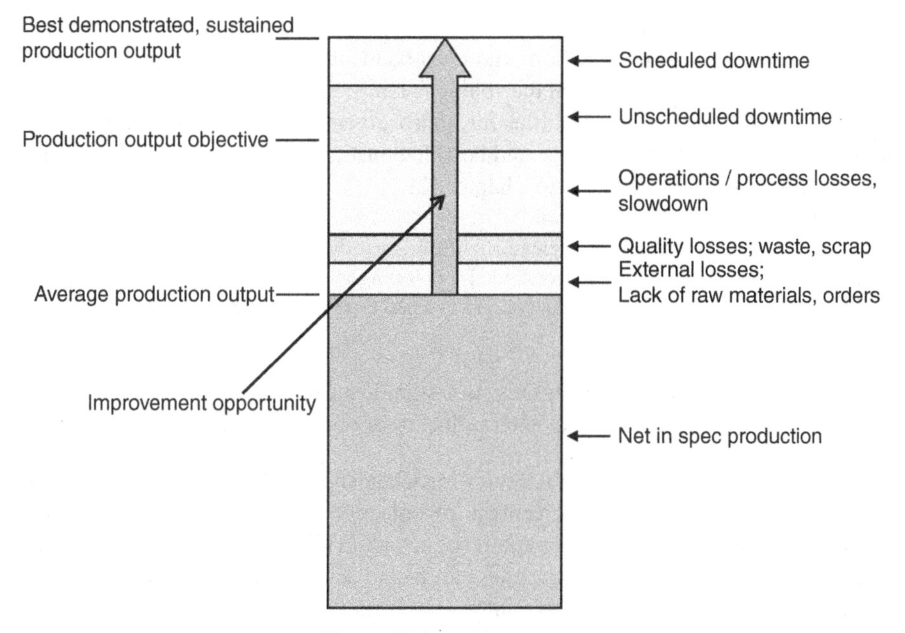

Figure 15.3 Hidden plant.

A facility processing perishable produce during harvest season operated only about 3 months a year. During the 3 months, operation at 100% availability was essential as the harvested crop (tomatoes) quickly spoiled.

An OEE below world-class values for the operation represents a "hidden plant." The "hidden plant" is the difference between best demonstrated, sustained production rate and the sum of operating/production losses that can be recovered by improved performance and effectiveness. Chapter 13 describes a method to determine the best-sustained production rate and the cause of lost production. As illustrated in Figure 15.3, the hidden plant includes losses due to the following:

- Downtime: scheduled and unscheduled
- Operations, process-related losses including set up, and adjustment time
- Quality and yield losses: waste and scrap
- External losses: lack of material and demand (sales)

In addition to operational deficiencies, production rate can be affected by sales or availability of raw material, which are largely outside the control of Operational Excellence. A lack of sales or raw material could force a reduction in rate that is neither a measure of the effectiveness of the operating unit nor is correction possible at the operating level. Unavailability, lost production, might be due to an operating problem, scheduled outage, and full or partial failure that can be corrected at the operating level; refer to Figure 15.3. In all cases, the specific difference is significant to determining improvement actions and must be identifiable in the data.

Awareness of the magnitude of the "hidden plant" and the specific causes provides the basis for identifying opportunities for improvement. Improvements are designed to narrow the gap between average and best-demonstrated production rate by attacking and eliminating losses; refer to Chapter 13.

One facility measured all three elements of OEE but did not multiply them together to create a single metric. They concluded monitoring availability, rate and quality, as individual metrics was more valuable and avoided a positive trend in one masking a negative trend in another.

The absence of a cost term in OEE is a significant weakness. As a result, OOE was introduced in Chapter 5 as a better indicator of operating effectiveness.

Overall Operating Effectiveness Reviewing Chapter 5, OOE is composed of availability, rate, and quality as a percentage of objective; the same as OEE, with the addition of cost objective as a percentage of actual. (This is not a typo—objective is in the numerator of the cost effectiveness term, and actual is the denominator.) Each element can be further broken down into component parts as in the hidden plant.

Lost Opportunity

Lost opportunity and the cause is another metric at this level. Lost opportunity represents lost revenue and can be caused by conditions such as a failure or partial failure, reduced efficiency, lack of raw material, etc; any condition that restricts production causing a loss; essentially the "hidden plant" in Figure 15.3. Lost opportunities can be especially costly and important if the loss impacts a customer or potential customer and cannot be made up in the future.

An operating organization experienced a partial failure that affected the economics of production. The nature of the failure presented a choice between producing a product fully meeting quality specifications unprofitably or defaulting on the order. Concluding that customer good will and continuing orders had greater value than temporary unprofitability, the production enterprise decided to deliver the order at a loss.

Quality

COPQ from Six Sigma is a strong metric that should be considered regardless of whether the enterprise is following Six Sigma. COPQ includes the full cost of all quality variations, including inspections, remanufacturing/scrap, penalties for missed deliveries, warranty cost, and lost customer good will.

Other Metrics

From here, there are many process specific metrics—reliability (MTBF mentioned earlier), maintenance work, and stores management to name three. For each of these examples, there are ample benchmarks published by professional societies to employ as objectives. If a large gap exists, the primary issue from the perspective of Operational Excellence is what should be the initial improvement objective that is seen as realistic and will energize efforts and create ownership and responsibility for improvement.

SELECTION OF METRICS

Applicability

The definition and calculation process for many metrics are identical and equally applicable across a variety of enterprises, facilities, and processes. Business, financial, some operational metrics such as OEE, and most maintenance work management metrics are in this category. Financial benchmarks such as RONA or EBIT are a valid standard of comparison across a variety of enterprises; values may be different. OEE has both similarities and differences. While the definition and method of calculation are the same across a wide variety of enterprises, standards of performance on the basis of OEE vary widely on the type of process from discrete manufacturing to continuous processing as mentioned earlier. Figure 15.4 broadly identifies metrics that are generally applicable, as well as those that are applicable to a specific industry or process.

As stated, mission, intensity of operations, location/environment, and age must be considered for an accurate comparison to performance benchmarks between enterprises and sites.

In a multi-business enterprise consisting of dissimilar operations, there may be broad variations in performance resulting from factors such as type of operation and intensity of processing. Leaders recognize that common factors to measure effectiveness are challenging to define when operations are significantly different. Under these circumstances, each operation must be evaluated on its own merits; performance in its specific environment. Here again, basing objectives on percentage improvement has advantages.

Best Measures

Results metrics that link performance, productivity, and effectiveness to business objectives and financial value should be emphasized. As stated earlier, metrics must be layered and linked top to bottom and bottom to top; refer to Figure 15.1. Figures 15.1 and 15.4 together provide a fair listing of the type and relationship of process and management metrics throughout the enterprise. They also indicate how working-level metrics link into site and enterprise performance and effectiveness metrics. Recognize that each layer must contribute demonstrable value to the layer above.

Metrics are developed as an interrelated family to assure alignment and avoid sub-optimization. Metrics in one area at one level in the hierarchy must complement and

	Metric
Corporate	Return on Net Assets (RONA) Return on Capital Employed/Invested (ROCE/I) Earnings Before Interest, Tax (EBIT) Total Rate of Return (TRR)
Industry Effectiveness	Cost as Percent of RAV / ERV / EDC Effective Forced Outage Rate (EFOR) Manufacturing/Production cost / Unit sold
Site / Facility Effectiveness	SHE Incident Rate Non Compliance Events Capacity Utilization
Business Unit / Product	Net Operating Profit Unit Contribution Margin Manufacturing/Production Cost / Unit Sold On Time Delivery
Operating Effectiveness	Run Rate, Throughput, Rolled Throughput Yield (RTY) First Run Quality (excludes rework) Overall Equipment Effectiveness (OEE/OOE) Cycle Time Cost Of Poor Quality (COPQ)
Asset Performance	Availability Mean Time Between Failure/Repair (MTBF/MTBR) Cost per Unit Output Surprise Failures
Maintenance Effectiveness	Uptime as a Percentage of Total Time Percentage Planned to Total Work Schedule Compliance Labor Utilization Overtime as a Percentage of Total Hours Repair success
Capital Effectiveness	Capacity Utilization Inventory Turns WIP

Figure 15.4 Applicable metrics.

strengthen metrics in other areas at the same level. The family must be scrutinized to assure improvement in one area doesn't detract from or suboptimize performance in another area. A rule of thumb is to assure that every metric reported upward in the hierarchy is fully supported by metrics at the level below.

Wherever possible, metrics must be expressed in terms that are either directly monetized or easily monetized. MTBF is an example. Normally expressed in months, improving MTBF is easily monetized by knowing the number in the population to be improved, objective results of improvement, and the average cost of repair; an example was cited in Chapter 12. Executives think and talk in monetary terms; it is the responsibility of leaders within the Operational Excellence program to position performance, improvements, and metrics in into monetary terms.

Concentrate on Results and Success

As stated, metrics, especially key performance indicators (KPIs) discussed in the next section should concentrate on value and demonstrate contribution to business results and operational improvements. Some recommendations for operations and maintenance working level results metrics are as follows:

Cost per unit produced—rather than production throughput

Program and task success; demonstrated value gain from improvement initiatives—rather than initiatives and tasks implemented or completed

Optimum system and equipment availability and lifetime cost—rather than system and equipment MTBF

Declining lifetime cost trends and failure rates—rather than Condition Based Maintenance (CBM) and Preventive Maintenance (PM) tasks completed

Declining surprises and unexpected failures—rather than percentage work originated by CBM and PM

Percentage root cause analysis (RCA) recommendations that succeeded in preventing subsequent events—rather than analyses completed

Repair success—percentage of repairs returned to Operations on-time, within planned hours, and restarted successfully with no follow on work required within 6 weeks, rather than rework.

KEY PERFORMANCE INDICATORS

KPIs are metrics selected to demonstrate performance to key strategic and operating objectives. KPIs best define an important aspect of effectiveness, monitor essential elements of the Operational Excellence program, and demonstrate progress to objectives. Total Recordable Rate (TRR) is a primary safety KPI. Cost per unit produced is a KPI of operating effectiveness. Production effectiveness is often expressed as OEE (OOE recommended as a better measure), uptime, or a similar industry-specific measure such as Effective Forced Outage Rate (EFOR) in the power generation industry. As stated earlier, many enterprises in the oil and petrochemical industries use cost as a percentage of RAV/ERV as the measure of cost effectiveness. There are many more.

KPIs are delivered to people with control responsibility. The process must include communication procedures to make certain the right information is getting to the right people in a timely manner. Processes must also be established for actions in response to KPIs; action to be taken by those responsible to assure the upward linkage accurately conveys performance and conditions in ample time to take corrective action in the event of discrepancies.

KPIs should be metrics in broad general use for which there is a clear definition, calculation procedure, and best-in-class benchmark values for reference comparison. KPIs are updated and published regularly to reinforce ownership and motivation.

> Several participants at a workshop commented their enterprises had too many KPIs. Furthermore, some KPIs were illogical and measured old activity-based behavior rather than effectiveness and results.

In general, there should be no more than three to five KPIs at each level of the hierarchy defined in Figures 15.1 and 15.4. Too few and all vital aspects of a given level within the enterprise may not be covered adequately. Too many without clear prioritization and KPIs lose their importance; people lose focus, and efforts are diluted. Each KPI must be within a clear line of sight and demonstrate contribution from level to level from bottom to top of the enterprise.

KPIs should be aligned with individual performance objectives.

GRAPHICAL DISPLAYS

Spider charts are frequently used to compare the performance of multiple metrics with benchmark objectives.

One large manufacturer plots audited performance of work teams in nine areas on a spider chart, as well as provides an overall assessment; refer to Figure 15.5.

1. Standards and documentation
2. People, organization, and culture
3. Costs and benchmarking
4. Communications and feedback
5. Training
6. Planning and scheduling
7. Planned maintenance
8. Supplies and tools
9. Continuous improvement.

A performance evaluation of each work team is conducted annually by members of other work teams as a performance assessment, learning, and benchmarking process.

The evaluation process begins by submitting written questions to a work team. The team provides written answers that are evaluated by those chosen to perform the assessment. The assessment team then conducts a 4-hour verbal review. The score in each area and the overall are plotted along with benchmark objectives to construct the spider chart shown in Figure 15.5. The spider chart provides a graphic means to identify gaps and areas of potential improvement. The spider chart and accompanying comments are combined into a feedback document that includes a written evaluation and suggestions for improvement. The latter is focused on improvement and success;

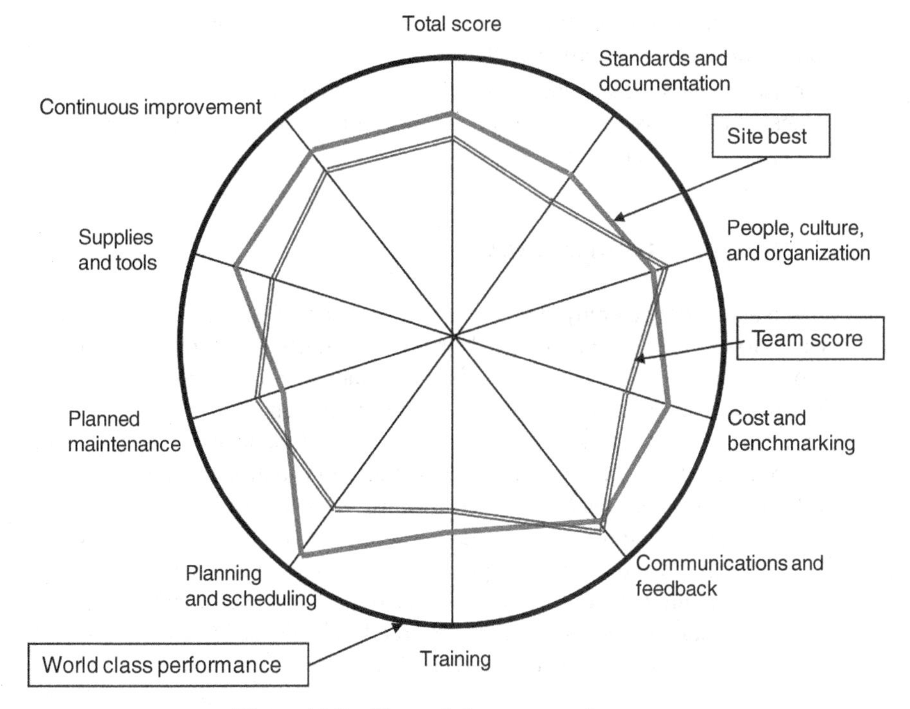

Figure 15.5 Chart of plant team performance.

not criticism. The team being evaluated has the option to apply improvement initiatives in areas other than those with the greatest gap on the spider chart. The team is accountable for results. Gaining the optimum mix of results for resources expended is the objective.

BENEFITS OF METRICS

Optimal metrics have multiple benefits. They convey the major value of an Operational Excellence program, identify the opportunity for and measure results of improvement initiatives, and create enthusiasm, ownership, and support. Specifics include documenting:

- Increased throughput demonstrating value gained from improved availability and decreased downtime, fewer outages, reduced slow time, and other delays
- Increased quality and reduced losses including waste/rework, returns, discounted sales, and penalties
- Increased energy efficiency demonstrated by energy costs before, during, and after improvements

- Reduction in costs: labor hours, parts, and consumable savings realized by minimizing failures and improving labor effectiveness
- Increased capital effectiveness gained by reduction in inventory and work in process made possible by greater predictability, reduced variation, reduced usage, and better planning

WHAT YOU SHOULD TAKE AWAY

Well-chosen and widely employed metrics are a vital part of Operational Excellence and essential to success. Beginning with the basis for identifying improvement opportunities to assuring objective value delivered and sustainable results, metrics are used continuously throughout the Operational Excellence program.

Metrics must be selected to accurately measure vital elements of performance and effectiveness. All must be linked to demonstrate contribution to enterprise strategy and business/mission objectives. Metrics must be objective and unambiguous. They must be relatively simple, easily understood, and communicated broadly. Where possible, the definition and calculation procedure of Operational Excellence metrics should be in full alignment with international standards and professional society recommendations. Data for metrics should be recorded and calculated automatically for consistency and to minimize time required and administrative burden.

Especially, important metrics representing key performance measures are selected and monitored as KPIs.

Performance to metrics and expectations must be publicized and communicated regularly to gain enthusiasm, ownership, and commitment to success.

16

IMPLEMENT—IMPROVEMENT ACTION PLANS

Once a comprehensive improvement action plan with clear objectives has been formulated and approved, the implementation process is "simply" executing the plan.

For maximum ownership and commitment, improvement action plans are implemented, managed, monitored, and adjusted by the Improvement Action Team, hereafter Action Team, which developed the plan augmented as necessary with specialized knowledge and skills. The team is responsible and accountable for results. Action teams begin with meetings at least weekly, or even at closer intervals where necessary, to assign specific responsibilities, assess progress, and identify any adjustments that might be required and/or unexpected barriers. Compliance to plan timelines and communications is exceptionally important to assure a fast start, increasing momentum, enthusiasm, and ownership. The team leader must push early and hard for results. Compliance with improvement milestones and results metrics must be top priority of all participants. As initiatives mature, action progresses, and problems are solved; meetings can be reduced to bi-weekly or even monthly. All involved must be alert for any slackening of effort or results. Early recognition and immediate correction are essential.

> Assembling a program is the easy part—getting it done is difficult! Excellence is not easy—if it was, it would be more crowded at the top!

Operational Excellence: Journey to Creating Sustainable Value, First Edition. John S. Mitchell.

Successful implementation depends on three additional factors: demonstrating contribution to business/mission objectives, communicating successes and results, and continuing high level support.

The Implement process is summarized in Figure 16.1.

REFINE THE ORGANIZATION

While implementing an Operational Excellence initiative; shifting to a multifunction, activity-based, work team organization may be considered; see Chapter 10. An activity-based team organization has the advantage of greater awareness, improved focus, better teamwork, prioritization, ownership, and efficiency. This approach makes sense when the function/crafts/trade "baggage" of the past can be discarded. Many operations people are skilled mechanics. Likewise, many skilled crafts do not hesitate to perform out-of-craft tasks on their homes or automobiles. The fact is that artificial boundaries by function, applicable only at work, limit efficiency—a condition that cannot be tolerated in today's climate. The challenge is how to make this type of organization perform most effectively.

> Years ago, a story surfaced about the implementation of a work order to lower paper towel dispensers in a restroom to meet statutory requirements for disabled persons. Implementation of the work order required a carpenter to remove the towel dispenser, a plasterer to fill the holes, painter to repaint, and the carpenter to reinstall the dispenser. Since multiple trades were involved, a supervisor was also required along with a driver because work rules forbade one craft from driving another to a worksite! Five people to perform a simple task that any one of the people involved would have performed without assistance if the requirement had been at home.

With its emphasis on multifunction Action Teams, Operational Excellence begins the process of improving teamwork and effectiveness within a functional organization without stressing the organization with potentially uncomfortable change that can significantly detract from the real objective. Within Operational Excellence, Action Teams are the bridge between operations, maintenance, engineering, and other functions, building the cooperative teamwork necessary to gain maximum results most effectively.

It has been stated that multifunction Action Teams often do better with leadership from Operations/Production. This is not a hard-and-fast rule; leadership from other functions can perform equally well—it is strictly dependent on the individual: commitment, ownership, and enthusiasm.

> In one large facility, action teams were led by a variety of people, including engineers, production, and maintenance managers. In every case, all involved knew that all the appropriate managers were fully supportive and behind the team and the team leaders. This empowered the leader, and all were successful.

Within the Action Team, individuals are selected and assigned to specific activities—generally by personal preference and willingness to become owners and

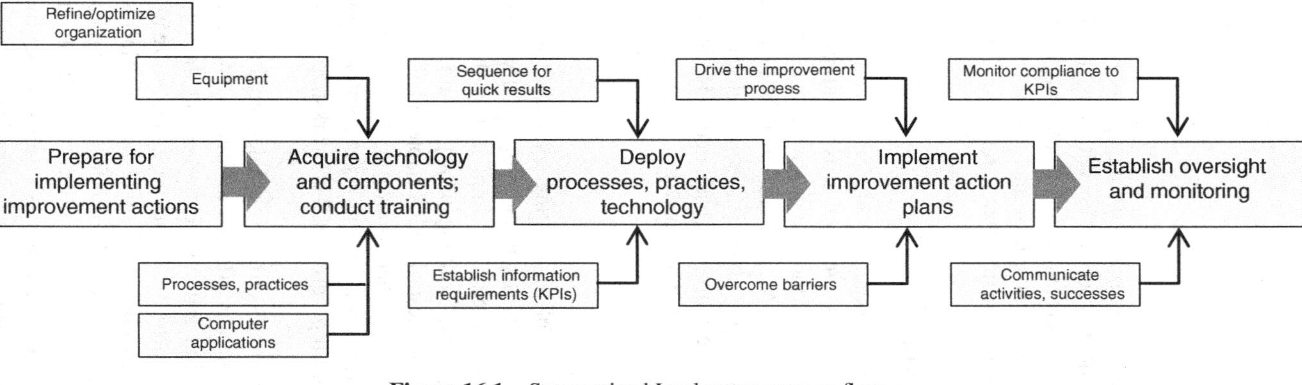

Figure 16.1 Summarized Implement process flow.

champions for specific elements of the improvement action plan. These individual owner champions select others as necessary to meet requirements and assure success.

DEPLOY RESOURCES

With the owner, champion leaders selected for improvement action plans, the next step is to begin the deployment process. This will consist of interlocking activities: purchasing capital equipment as required and conducting training to assure the plan is deployed efficiently; processes, practices, and equipment called for in the plan are used effectively. At this point the action plan is ready for full implementation.

Purchase Capital Equipment

If the action plan implementation includes capital expenditures for instrumentation, computers, computer applications, etc. and/or expenditures for external facilitation and training such as mobile applications, Lean, Root Cause Analysis (RCA), and Failure Modes Effects Analysis (FMEA), they are generally identified and costed during the identification and planning processes. If specific selections have not been made, requirements are identified, alternatives investigated, selections made, and purchased at the beginning of the implementation process.

Conduct Training

There are several types and levels of training conducted at the outset of the Operational Excellence initiative. Training in the program itself and specific training to assure technology, processes, and practices implemented during the program are used most effectively are two key areas. Training in the program itself is best conducted in a workshop format similar to that described in Chapter 11. Training continues throughout the program as necessary. *As necessary* might mean continuing program and awareness training along with communications to reinforce principles and broaden lessons learned as the program progresses. The workshop format is highly effective for training.

Communications and awareness training are delivered as appropriate to all in the enterprise, as detailed in Chapter 10. The awareness and training process should begin as soon as the decision is made to implement the Operational Excellence program in order to build enthusiasm and support—minimize misunderstandings and rumors. Awareness training is conducted on at least two levels: one for executives and senior management and the second for informative purposes for people within the organization who are largely uninvolved with implementation. Awareness training includes program description, necessity, objectives, principles, work culture, values, and, of course, benefits. Awareness training is best conducted by the program leader/champion who will benefit significantly from the inevitable questions and workforce concerns that must be addressed to gain acceptance and cooperation. It is a good idea for a member or members of the Communications Team to participate in awareness training for the same reason. Communicating questions and concerns of broad interest that are discussed and answered during training is an excellent method to assure everyone has valid information and adds to the body of information needed for enthusiastic support.

Detailed program training is developed for those directly involved to assure knowledge necessary to achieve success. It begins with workshops at the initial stages of the program, as explained in Chapter 11, and continues with specific reinforcing knowledge as required.

There will be multiple types and variations of training in technology, systems, processes, and best practices conducted as part of the Operational Excellence program. RCA is a good example of the scope of technology and practice training, as it is conducted in three levels: training must be based on a fully documented procedure that specifies application and detailed methodology. Detailed process training is required for people who will be leading RCA teams. This is normally conducted by an outside trainer, typically an experienced expert from the supplier of the RCA process, who may remain as a facilitator during deployment and initial use. Next, people who will be members of RCA teams are trained in the procedure. This is typically less detailed and delivered by RCA team leaders. Both the team leader and member training will include working through actual examples of a RCA. Finally, all who may be involved must have awareness training for an understanding of the process, its benefits, and requirements. The latter includes the essential nature of recording and preserving vital data needed for the accurate, detailed analysis and resulting recommendations necessary to assure incidents don't repeat.

As a final comment before going on to the next stage—becoming a trainer has major advantages for members of the Operational Excellence implementing team. It requires the person delivering training to become a real expert, as participants quickly see through any weaknesses or shortcomings in knowledge. A second major benefit is the back and forth between trainer and trainees during the process. Often, excellent questions are asked that the trainer, no matter how experienced, hasn't a good or refined answer. The training process is educational for the trainer and trainees alike—ask any experienced trainer!

DEPLOY PRACTICES AND TECHNOLOGY

Practices and technology, such as Lean, Lean Six Sigma, RCA, RCM (Reliability-Centered Maintenance)/FMEA, TPM (Total Productive Maintenance), PM (Preventive Maintenance), and CBM (Condition-Based Maintenance)/PdM (Predictive Maintenance), are deployed in accordance with the improvement plan. Recognize that within Operational Excellence, practices are deployed to address specific opportunities, not necessarily on an overall, broad basis as is customary; refer to Chapter 4. Practices in place before the Operational Excellence initiative and determined in need of improvement during the identification and planning processes are reviewed and upgraded as necessary to gain optimal effectiveness and maximum value.

Here again, it is wise to consider the use of an experienced third-party facilitator to get off to the best start. A third party facilitator can provide major benefits implementing new practices and/or technology as well as improving in-place practices and technology. In both cases, a third party can move the process of implementation/improvement faster and with much greater confidence and reduced possibilities of a misstep or omission that may cause significant problems later.

IMPLEMENT IMPROVEMENT ACTION PLANS

Improvement action plans are implemented by teams that, it's assumed, gained full commitment and ownership for success during the planning process and are now willing to be accountable for results. Success requires participation, ownership, and above all teamwork and cooperation. With the latter, challenges are quickly identified and overcome.

Implementation should prioritize quick success gained by harvesting "low hanging fruit." The "quick wins," known problems with actions for rapid correction, should have been identified in the action plan. Nothing increases motivation, enthusiasm, confidence, and support better and faster than successfully quickly solving real problems!

A major improvement process at a large site began with a month of classroom training, followed by several months to broadly apply the training. The process failed in less than 6 months. When participants were asked (some volunteered), they stated it was impossible to maintain interest or motivation necessary, or justify spending time, to continue with a process that didn't appear to contribute anything significant to solving specific problems they had to deal with on a daily basis.

As the program progresses, there will be multiple improvement plans in various stages of maturity. Coordination is essential to assure optimal overall results. Coordination must begin at the working level where action team members are encouraged to communicate and share experience with team members working on other initiatives. Coordination at the upper levels occurs during Leadership and Steering Team meetings. The overall message is awareness so that cooperation improves results and minimizes conflicts. As lessons are learned, they need to be disseminated widely through the training process.

Recognize that additional resources may be required during implementation. People with specialized knowledge and computer applications are two examples.

An improvement team pursuing several initiatives to increase plant OEE (Overall Equipment Effectiveness) found a convergence between what had been assumed were independent initiatives. To investigate this further, perhaps identify a core improvement, the team enlisted assistance from several people with specialized knowledge. Within a short period, linkages had been established that enabled consolidating several improvement initiatives into one, thereby streamlining the improvement process and improving the probability of success.

As action teams are engaged in implementation, the Program Leader, Steering and Communications Teams are working in parallel to drive the overall organization and cultural transition that is essential for the "new way" to take root and flourish; refer to Chapter 10. Examples are the best teachers. Since action teams were chosen for enthusiastic commitment and ownership, examples will quickly emerge during the implementation process that will reinforce the necessity of changing the work culture from status quo to "how can we make conditions better." Examples are factored into the training program and communicated widely.

At the opposite end of the scale, all involved must be sensitive to cultural and organizational barriers, as well as identifying any and all sources of friction. Positive steps must be taken quickly to minimize, preferably eliminate, counterproductive attitudes, and behaviors.

In the "real world," some highly experienced and very valuable people are simply not well suited to work effectively in teams. When this occurs, the program leader and associated management must make every effort to use the experience, skills, and potential contribution of these individuals to gain maximum value. Using them as internal consultants to solve specific problems or in special, complementary projects that can be accomplished alone are two possibilities. Both have been used successfully to gain maximum individual contribution.

> A highly competent and experienced individual simply wasn't suited to working in a team. He quickly became impatient with individuals who were typically much younger, didn't have his level of expertise, didn't grasp details as quickly, and appeared to him less motivated to perform all work with the utmost quality. To use his exceptional motivation, knowledge, and skill most effectively, he was assigned special projects where he could work alone to solve particularly difficult problems. He excelled in this role providing great value to the facility.

ESTABLISH INTERNAL OVERSIGHT AND MONITORING

As improvement action plans are implemented and gather momentum, it is imperative to establish internal monitoring to reach and maintain schedule. Facilitation, coaching, and assistance must be applied quickly when challenges arise and/or schedule slips. Since schedule has a large impact on value gained and program support, especially for "quick wins," every effort must be made to comply. Weak areas and improvements falling behind schedule must be identified and corrected as soon as possible.

> At a large site, management allowed the schedule for improvement initiatives to slip with variations on the comment "didn't have time this month, will regain schedule next month." Months passed with continuing delays, until insufficient time remaining forced management to reveal that promised year-end objectives couldn't be met. In another, similar incident, a year's delay to a major, multi-year project was announced about 6 weeks before scheduled completion. Predictable last minute announcements such as these are horribly discrediting to a program and can't be allowed to occur. Progress

milestones must be developed and adhered to. Strong, positive corrective action must be taken at the first indication of a deviation from plan.

Particular attention must be given to demonstrating value contribution to business/mission objectives. The question: "how does this activity/task contribute to business/mission objectives" must be kept uppermost in mind throughout the process.

DRIVE THE IMPROVEMENT PROCESS

The Executive Champion, Plant Manager, Steering Team, Program Leader, and all initiative champions must continually press for action and results. Their visible, continuing presence, interest, and support are essential to reinforce the relevance and importance of the program and necessity for results. In many cases, people directly involved with implementing improvement initiatives will have so many demands on their time that today's crisis tends to take precedence over anything that can be accomplished tomorrow. Challenge is that without discipline, work toward future improvement is always tomorrow! Senior executives and management cannot expect people to make a commitment to a difficult initiative if all they do is say some good words at initiation and then disappear. Continuing visible support from executives and management is imperative for success.

Everyone must understand that time invested in Operational Excellence is essential to continuing success and is well spent. Without investment in improvement, tomorrow and the next day will be just like today!

OVERCOME BARRIERS

Barriers to success will be encountered, some predictable and some not. In some cases, the barrier may be procedural or operational that can't be eliminated without a major change.

Organizational and personal conflicts can generally be resolved. Rules to avoid conflict must be established early in the improvement program and enforced. The basic rule is one of values; people and their ideas are always treated with respect. Facilitators, leadership, and everyone in charge of and participating in meetings and discussions must be sensitive to this issue and stop infringement immediately.

Anticipating barriers, personnel implementing an improvement initiative developed parallel paths so that if one was blocked for any reason, activities could be shifted to a second initiative without any loss of momentum while the barrier was being addressed and resolved.

The availability of time is often the greatest barrier of all. To surmount, people must be convinced that time invested in improvements today will be returned tomorrow with greater order, less variation, and fewer problems.

At launch of a major improvement program, a site reorganized and reassigned tasks so that improvement team members could devote a minimum of 50% of their time to the process. The realignment had a number of significant advantages: the improvement initiative benefited and was significantly strengthened by the participation of the most experienced employees with greatest insight into the company organization, strengths, and weaknesses. Less experienced employees had the opportunity to step up into greater responsibility with assurance of immediately available support in the event of uncertainty or questions.

Overcoming some barriers, particularly conflicts within a team, may require individual counseling or even more general training. In-depth team training may be useful at the beginning of the improvement initiative so that everyone understands the principles and etiquette of working productively together in a team, is familiar with the mission, and can quickly focus on the tasks.

COMMUNICATE RESULTS AND SUCCESSES

As has been stated earlier, continuing communications are all important. Communicating results and successes builds enthusiasm and ownership and contributes greatly to the overall support needed for success. Feedback from suggestions for improvements to operations and the program itself are particularly important and should include actions taken. Partial results, what people are doing to identify and solve problems, also should be communicated to demonstrate the serious commitment to success and encourage wider participation.

WHAT YOU SHOULD TAKE AWAY

Refining multifunction Improvement Action Teams to assure all the experience and knowledge required to assure success are resident within the team is the first step during implementation. Furthermore, emphasizing cross-function teamwork though the action teams establishes the model for constructive cooperation that is essential throughout the organization.

Training is conducted to assure awareness, support, and the detailed knowledge necessary for acceptance, ownership, and success as resources are deployed to implement improvement plans.

The information structure and data storage are reviewed to assure data are available to identify and value improvements, that results will be demonstrable, and additional opportunities will be identified.

Full compliance with interim progress (activity) and final (results) KPIs (Key Performance Indicators) is essential. Any departures must be recognized quickly, investigated, and corrected as soon as possible.

Results and successes are communicated widely to build enthusiasm, support, participation, and success.

17

PERFORMANCE ASSESSMENTS

Assessments provide a standardized, disciplined method for evaluating performance, effectiveness, and results achieved by the Operational Excellence program. Assessments can be directed to the overall program, a function within the program (maintenance), a process (reliability), or a procedure (Root Cause Analysis) at one point in time. Assessments establish the current status of the program, people, processes, and technology. They determine if the program is fully defined, chartered, and is producing results; if people are properly trained; if there are adequate resources and work processes to support the improvement effort; and if existing technology is being fully integrated and managed.

Periodic performance assessments are required by ISO55000.

The assessment process described in this chapter begins with detailed objectives determined from benchmarks and best practices, includes criteria to assure maximum objectivity, and concludes with formal conclusions and recommendations.

Performance Assessments are used for three primary purposes as follows:

- At the beginning of an improvement initiative to establish and accurately document initial performance and conditions. The assessment is structured to identify and provide a comprehensive list of improvement opportunities, with full description of best practice objectives. This highly important use assures a new improvement program is focused into the areas where results and value can be achieved most rapidly with least effect.

Operational Excellence: Journey to Creating Sustainable Value, First Edition. John S. Mitchell.

- During the implementation of improvement initiatives to provide an accurate, objective, and consistent means to periodically evaluate performance. The assessment includes monitoring a program or activity's progress to objectives, overall and specific effectiveness, compliance with specific time milestones, and results. It also includes identifying trends from previous assessments and additional opportunities for improvement (gaps) in one or more areas of specific activity. The assessment identifies and tracks progress in a formal, disciplined, and auditable manner and provides the basis for developing additional improvements to elevate performance.
- Periodically at regular intervals (typically not to exceed 18 months) to measure performance and effectiveness, assure sustaining compliance to best practices, and identify additional areas for improvement.

OVERALL DESCRIPTION

Performance-based assessments begin by identifying the program or function to be evaluated.

A formal compilation of all elements necessary for success—performance objectives, including description of best practice/benchmark requirements are assembled. These serve as a template of comparative standards to measure effectiveness and results in the areas to be examined. A template improves speed and effectiveness and assures all areas vital to success are considered. Actual performance, effectiveness, and results in the selected areas are documented in detail through reviews of procedures, records, and interviews. Actual performance is compared against standards using objective criteria to identify opportunities for improvement (gaps). The comparative analysis between current performance and best practice leads directly to recommendations for improvement.

It is essential that all personnel in the areas to be assessed understand that the assessment is a learning process, strictly directed to identify the performance, results, and effectiveness of current methods and practices, as well as to identify opportunities for improvement. It must be stressed before and during the process that the assessment is not performed to criticize or assign blame.

ASSESSMENT METHODS

There are two methods to evaluate performance. A subjective assessment is essentially an individual or group opinion. Examples include restaurant/movie reviews; the former includes ambiance (layout, lighting, noise level, décor, table setting, etc.), service, food quality, price/value, and overall experience; all based on opinion.

An objective assessment is actual performance compared to defined rules and criteria. Golf is an example where performance and number of strokes are compared to course standards, par. Athletic standings (competition against others) is a second example. A school report card is a third example where a letter or numeric grade; A, 4.0, etc. compares performance to knowledge—knowledge of course content determined by periodic and final testing. The grade may also include homework completion and possibly classroom conduct, the former an activity metric; refer to Chapter 15.

A detailed assessment template combines the two concepts to provide objective grading criteria for areas in which performance may be highly subjective. A detailed template identifies specific requirements to attain best practice performance along with comparative standards; for example, adding objectivity to the restaurant assessment cited previously.

ASSESSMENT PROCESS

A successful assessment has the following basic steps:

- Preparation:
  - Identify facility and operating unit to be assessed
  - Identify the specific program, program element, and procedure to be assessed
  - Identify all stakeholders
  - Select a qualified assessment team and team leader
  - Strategically plan the assessment to include the following:
    - Define specific objectives, areas, and elements to be incorporated in the assessment
    - Identify benchmark and best practice performance—results and effectiveness objectives for each element of the assessment
    - Detail objective comparison criteria
    - Establish timing—commencement and completion
    - Conduct preassessment team training as required
    - Conduct preassessment self-evaluation by entity to be assessed
    - Align the assessment team and assign responsibility for specific areas of the assessment
- Conduct performance-based assessment
- Compile results
- Perform opportunity (gap) analysis and develop recommendations for improvement
- Submit preliminary report, including recommendations for improvement —discuss findings
- Prepare and present final report
- Manage and support improvement actions.

ASSESSMENT PREPARATION

Identify Operating/Functional Unit and Requirements

An operating/functional unit will either request an assessment in a specific area or will be scheduled for a regular, periodic assessment to validate performance and identify opportunities for improvement.

Identify the Specific Program, Program Element, and Procedure to be Assessed

Requirements will establish the scope, content, and boundaries of the assessment. This leads directly to identifying stakeholders and qualifications required on the assessment team.

Identify Stakeholders

There are typically the following four primary stakeholders to be consulted and involved in an assessment process:

- Operating unit sponsoring leadership
- Working-level leadership in the specific areas to be examined
- Operating unit/functional personnel performing the activities to be examined
- Group performing the assessment: assessment team.

Select a Qualified Assessment Team and Team Leader

An assessment team normally consists of four to six people: leader and subject matter experts with detailed, complementary knowledge of the areas to be assessed. The assessment team may be augmented for specific requirements (operations, engineering, IT, finance, and HR, as applicable). Team members must have experience and/or training in the assessment methodology and process.

A team leader—experienced with assessments and the assessment process and preferably leading, is appointed to direct, coordinate, and manage the process. The team leader is responsible for establishing all requirements for the assessment, including scope, the process itself, team members responsible for interviews, and the interview schedule.

The team leader is responsible for compiling the information gained during the assessment, interim, and final reports.

Site Appoints Host and Establishes Timing, Commencement, and Completion

The operating unit being assessed appoints a host for the assessment. The host will be responsible for reviewing the assessment process and template, assuring processes and terminology are aligned, accomplishing the preassessment self-evaluation, and coordinating the timing of the assessment with the assessment team leader. The host arranges the assessment interview schedule: pre- and post-assessment discussions. The host performs introductions to interviewees, as well as obtains supporting documentation.

Strategically Plan the Assessment

Define Objectives, Scope, Areas, and Elements to be Incorporated in the Assessment Objectives, scope, and areas to be incorporated in an assessment are normally defined by the specific requirement and/or schedule. Examples are assess the effectiveness of the reliability and/or maintenance process, a specific element within the process such as Preventive Maintenance or work management, or even a segment within a process, for example, planning.

Identify Best Practice Performance, Results, and Effectiveness Objectives It is essential to establish standards from which actual performance will be compared for each element of the assessment. Standards of performance may be published benchmarks and/or descriptions of best practices obtained from other sources and/or improvement objectives from a previous assessment. All must be documented; an assessment template format is recommended. An extract is shown in Figures 17.1 and 17.2.

Detail Objective Comparison Criteria It is essential to have a detailed list of specific questions that will be asked to evaluate comparative performance. An assessment template, such as that summarized in Figures 17.1 and 17.2, with concise, comparative grading criteria is advantageous to assure everyone in the assessment team has a

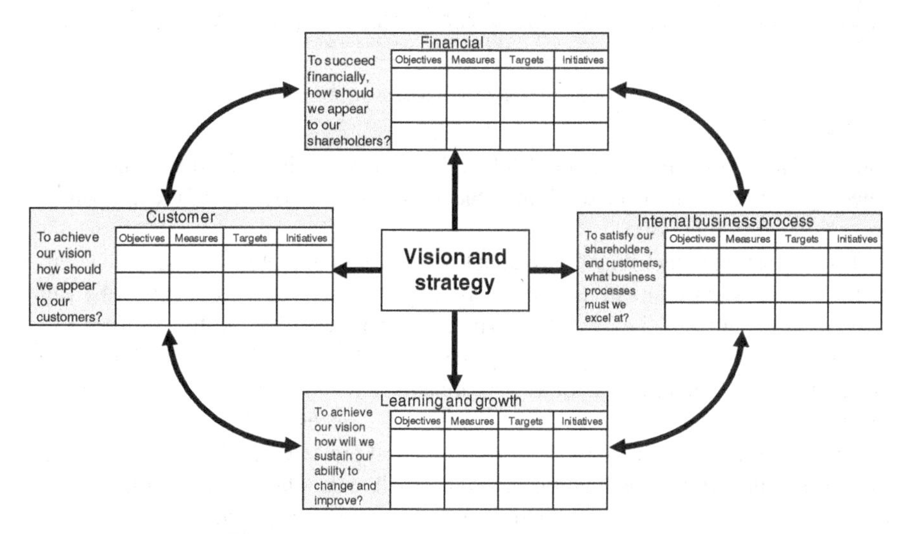

Figure 17.1 Example of a Balanced Scorecard.

Risk, Identification, Assessment, and control	Industry best practice
Risk of adverse events that could potentially occur on systems and assets. **Formalized, standard method fully documented, in place and utilized to calculate and determine risk: probability times consequences for systems and assets**	■ Risk, risk minimization and control is a primary driving force at the site. ■ Risk rank assessment results drive specific techniques for determining equipment strategies; strategy application and repair priorities ■ Major, formalized effort made to proactively identify and minimize risk of events that have never occurred ■ KPI's/measures in place to drive and determine effectiveness of risk ranking system.
■ Process(es) and/or procedure(s) for fully identifying and assessing risk, including analysis, determination/ranking, management and implementation of control/mitigation measures established, documented, implemented and maintained up to date	Robust, documented, and well-understood process for risk ranking assessment of assets and systems in place and followed.
Risk assessment process ■ Based on the probability of an event and all consequence including: a) Compliance with statutory, regulatory, HSE requirements to include cost of incidents, fines b) Physical failure such as functional or hidden failure, incidental damage, malicious or other action c) Operational risks, including meeting all performance expections, loss of control, human error and factors that might affect HSE, performance or condition d) Natural environmental events; storms, flooding, etc. e) External, imposed conditions; interruption in external services such as electric power, water, waste disposal f) Business risks such as failure to meet delivery commitments or reputation damage g) Risks during desigh, specification, procurement, construction, installation and decommissioning	■ Risks with high consequences, low probability (black swan) subject to rigorous assessment ■ Methodology includes provisions to level or normalize risk for consistency across systems and the organization ■ Identification, assessment, control and mitigation methods appropriate to magnitude of risk ■ Assessments prioritized by risk and accomplished proactively (before rather than after an event)
■ **Assessments to include:** a) How risks may change with conditions or time b) Risks inherent in adequacy of knowledge and / or assumptions utilized in the assessment c) Probability that control and / or mitigating actions will effectively reduce risk	All conditions fully considered in assessment
■ Calculation—Specific method defined and utilized to numerically quantify risk.	Detailed calculational methodology fully documented and utilized
■ Risk classification and prioritization—Risk classified/prioritized by categories; e.g., high, medium, low. Portion of the total systems and assets classified highest risk?	Risk classification defined and fully utilized; no more than 15% of systems and assets in highest risk category Full understanding for establishing the 'right' surveillance and management at the 'right' level based on risk
■ Records—Complete assessments including all assumptions, logic, conclusions, control and mitigation recommendations fully documented and maintained as permanent records	Full, auditable records established and maintained
■ Process review—Risk ranking process; including assumptions, assessments and results, reviewed when changes (conditions, regulations/requirements, operations, etc.) occur and at regular intervals. to validate conditions, assumptions, logic, conclusions and effectiveness of actions; updated as required	Reviews conducted when changes occur and periodically (no greater than 3 years) to assure currency; incorporation of any changes due to changes in regulations, raw materials, operation, equipment condition, etc.
■ Audits—Audit trail established and supported that documents the entire logic and process	Audit process formally established and proactively implemented to assure auditable compliance with all legal, regulatory and site risk control procedures

Figure 17.2 Extract from assessment template.

common basis for their evaluation. With performance, results, and effectiveness evaluated against a common standard, the objectivity and consistency of the assessment are significantly increased.

The assessment team may either adapt an existing template or create a new template. If an existing template is to be used, questions and terminology must be validated by the team and site host. This will assure applicability and full conformance with systems and terminology used by the site to be assessed. Terminology, categories/domains, elements, and questions are adjusted and refined as required to gain full alignment. Levels of performance are adjusted as necessary to maximize objectivity and consistency, assure valid comparison to best practice benchmarks, and clarify any areas of uncertainty. Validation should normally take place between 1 month and 2 weeks before commencing the assessment. The assessment template is tested and further refined following the preassessment self-evaluation.

If a new template is to be prepared, cooperative work should begin between the assessment team, site, and possibly a site team at least 1 month in advance of commencing the assessment.

In either case, it is crucial to obtain agreement between the site and assessment team regarding scope and specifics before commencement of the assessment.

Establish Timing: Commencement and Completion Timing, commencement, and completion are established cooperatively by the team leader and the operating unit host, considering availability and schedules of the assessment team and people at the operating unit to be interviewed.

Conduct Assessment Team Training Assessment team training is conducted as required to assure all team members are familiar with the process, specific areas to be investigated, and performance criteria on which the operating unit will be compared.

Operating Unit Conducts Preassessment Self-Evaluation A self-evaluation questionnaire is distributed to participants within areas at the operating unit to be assessed. Participants are asked to evaluate elements included in the formal assessment in the same form as will be used in the assessment. In this way, the assessment team has an idea of how the operating unit ranks its own performance. The self-evaluation has a second purpose—to validate the assessment template for the operating unit's specific requirements, activities, and terminology. The assessment team leader may or may not include grading criteria in the preassessment questionnaire.

Responses are scored by the assessment team and plotted on the diagram that will be used for the assessment itself (i.e., histogram and spider plot). A comparison between the self-assessment and the assessment team's consensus and reasons for any differences is, in itself, a valuable learning tool.

Align the Assessment Team The assessment team meets before commencement to assure objectives are fully understood, important questions and information to be discovered are identified, and responsibilities fully defined. The following specifics must be addressed before commencement:

- Define responsibilities for specific areas of the assessment
- Identify documentation required from the operating unit to be requested from the host
- Review the operating unit self-assessment and identify specific areas and questions to probe
- Validate the assessment scheduling and interviewee list. Review interview schedule and assure at least two persons are assigned each interview
- Define responsibilities for notes—observations, conclusions, recommendations, and final report
- Discuss and finalize criteria for numerical grading

- Discuss the probability that some individuals to be interviewed may feel threatened and defensive. Discuss ways to turn suspicion, reluctance to answer questions, into a positive learning experience (sensitivity to protecting candid observations).

PERFORM THE ASSESSMENT

The operating unit assessment is conducted in four stages:

- Initial alignment meeting
- Assessment
- Conclusion, summary, and wrap up
- Preparation and submission of final report

Alignment Meeting

The initial preassessment alignment meeting occurs when the assessment team first arrives on site. Operating unit personnel who should be in attendance for all or part of the meeting include operating unit sponsor, senior operating leadership (optional—up to the facility/site/unit), assessment host, assessment team, and representatives of the people who will be directly involved.

The meeting is generally arranged in segments as follows:

- An overview of the assessment centered on purpose, results, and actions required. This is primarily for operating leadership to assure full understanding of the process and expectations from both the operating unit and the assessment team are aligned. The alignment meeting also serves to obtain endorsement and support from the operating unit leadership.
- A summary for people directly involved to assure all understand the procedure, objectives are aligned, and any questions answered. The schedule, assessment template, and interview list are reviewed and validated.

It is also necessary to make certain there is a mutual understanding that an assessment is a learning experience to identify opportunities for improvement—not a punitive exercise to identify deficiencies and place blame.

If any safety briefings are required for the assessment team, they are conducted during this initial stage.

Assessment Procedure

An assessment generally requires 3–4 days for a full operating unit assessment; proportionally less for a specific function or process.

Occasionally, the assessment team must recognize that some people being interviewed may not be fully briefed on the purpose and, as a result, react defensively.

Interviewers have about 5–10 min to allay suspicions. Interviewers must be able to quickly recognize and ease a potentially hostile situation.

> At the beginning of an assessment interview, a site individual stated with arms folded that he didn't know the purpose or why he was there. Furthermore, he resented outsiders questioning him on his methods for carrying out responsibilities and didn't have any time to spare. Terrific start to an interview that ended very well!

Where practicable, it is best to conduct most interviews at the interviewees work area. This provides an indication of real job responsibilities, efficiency, and interruptions. The person being interviewed will normally be more comfortable in familiar surroundings and may want to emphasize or clarify an answer with an actual example. Visual displays, performance charts, and surroundings are valuable indicators of interest, commitment, and performance. Finally, a constantly ringing telephone or people popping in with questions say a great deal about procedure and organization.

> An assessment interview was constantly interrupted by telephone calls and people popping in with questions. When asked, the interviewee stated that this was a typical day. Although many of the questions could be answered from readily available references, people didn't appear to want to bother.

Site Walking Tour At some point during the assessment process, a walking tour of the site at the working level (more than a drive through) is highly recommended. In an overall assessment, orderliness, cleanliness, and safety precautions are powerful indicators of performance, commitment, and ownership. Checking calibration records, dated inspection tags on fire, and rigging equipment is very revealing.

> Before arriving on site, it was made clear to hosts that a working-level walking tour was essential. The particular plant was set up as a showcase with a visitors viewing gallery above the working area. The initial "walking tour" was conducted by a company tour guide through the visitors gallery following a bus load of high school students. When the site host was informed of the necessity for a working-level tour with a shift operator or equivalent to assess actual conditions, the response was a tour of the warehouse. The individual responsible for the warehouse was so proud of the organization, orderliness and cleanliness, and rightfully so, it was difficult to state that the tour still didn't meet what was needed. Walking outside the warehouse, the site host stated that "your business dress isn't proper for a working-level tour." When asked what was missing, he stated steel toed shoes. When the visitor proved he was wearing steel toed dress shoes, the host finally organized a full working-level tour including the control room.

Moral of the preceding story is that all assessment team members must come fully prepared to visit the facility, including all Personal Protective Equipment (PPE) that might be required. Hardhats are one exception; they are generally provided by the facility.

Debriefs It is quite valuable to have a discussion and debrief of the assessment team and the operating unit host after each day's interviews and activities. The discussion includes an information exchange, summary of findings, and review of the next day's schedule. The discussion should also include any additional questions and areas to investigate on the basis of the day's findings.

The next to last half day should be kept free for team discussion (debrief) and preparation of the departure summary report to be delivered on the final day at the operating unit.

ASSESSMENT TEMPLATE

Background

The idea of a scorecard on which to evaluate operating activity has been present for decades. The Balanced Scorecard®, shown in Figure 17.1, is an example of a process in common use. Within the Balanced Scorecard shown in Figure 17.1, requirements are defined in four categories as follows:

- Financial
- Internal business processes
- Learning and growth
- Customer.

The Operational Excellence assessment template, mentioned in this section and illustrated by an extract in Figure 17.2, began development in 2003 as a comprehensive reliability program scorecard. The scorecard was developed as a listing of all elements necessary for a complete reliability program and included summarized best practice and objective grading criteria to compare actual performance with best practice. The original scorecard also included a system of weighting to account for differing risk and potential value of specific elements within the scorecard. The original scorecard received favorable response during several conferences in the United States and Middle East.

The 2003 scorecard was refined during four 90-min workshops sponsored by reliabilityweb.com during the IMC conference, December 2005. Each element was reviewed in detail with a group of experienced people who made numerous constructive comments and suggestions. After the workshops, a second, revised version was circulated in early 2006. The revised scorecard had expanded to twelve program categories. It included areas such as results, organization, and supporting processes, as well as numerous improvements based on comments and suggestions during the workshops.

The concept of weighting by importance proved daunting. The workshops demonstrated that both risk and importance were heavily dependent on the type of business/mission, site/facility, processes, and program objectives. It was difficult to generalize, as there were large variations between industries and processes. With

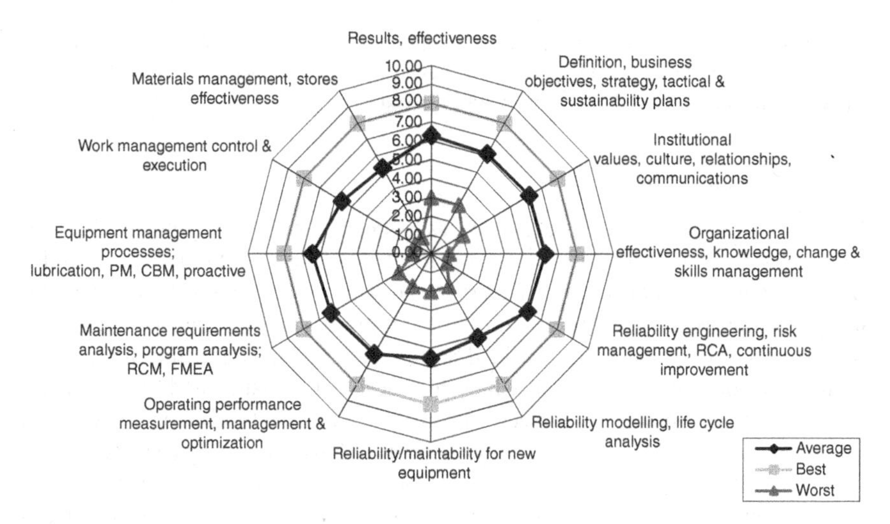

Figure 17.3 Assessment results radar plot.

an inability to gain any sort of general consensus that would be applicable to all who might use the scorecard, the idea of weighting was dropped. It should be reconsidered by any who contemplate using the scorecard for a specific operating, site, or process.

The 12-category/domain format led to recommendations for a standard reporting format. Numerical grades for each category (dimensionless numerical grades representing actual compared to best practice) would be displayed graphically in a spider/radar plot with results/effectiveness at 12-O'clock. Figure 17.3 illustrates an example. Remaining categories are arranged in a logical sequence clockwise. Later discussions demonstrated that everyone has ideas of how to best display results; the spider/radar display is only one of the many.

The revised assessment scorecard was published on the Internet and in the September-October 2006 issue of *Reliability* magazine.

In 2008, the reliability scorecard was extended to physical asset management. Results, supporting practices, and procedures were expanded to reflect the increased scope. By this time, the scorecard had been adapted for use by several operating units. All reported that the scorecard was an excellent starting point that had resulted in successful assessments.

In 2010, the scorecard was expanded again for applicability to Operational Excellence. As a final modification, the name assessment scorecard was changed to a more friendly assessment template. An operating unit reported back on the process used and success achieved with the scorecard.

> One major, multisite enterprise assembled an assessment team and used the template as a starting point. It required approximately 1 week to fully align the contents with their practice and terminology. The customized template was then used very successfully for

a number of facility assessments. The individual heading the effort stated that beginning with a well-developed template was a huge time saver and led to highly effective assessments that were well received by all involved.

Template Organization

The current Operational Excellence assessment template, refined for an operating/production enterprise, is organized into the eight natural affinity groups/categories introduced in Chapters 3 and 4. Each contains the elements that must be considered to meet business/mission, program, and performance objectives. Full descriptive information is provided so that requirements are fully understood and there are objective criteria on which to assess performance and effectiveness. Categories are organized by subject and groups within a typical operating unit. They are neither equal sized nor do they necessarily have equal emphasis. Emphasis is determined by specific assessment goals of the site. In many cases, it may be more efficient to subdivide a category, practices, and procedures as one example, to concentrate attention on particularly vital processes, practices, and procedures.

The eight categories are as follows:

1. Overall results: Safety, Health, Environmental (SHE) performance and contribution to enterprise operating/mission objectives.
2. Leadership at the executive and operating levels.
3. Requirements: SHE, legal, statutory and regulatory, insurance, contribution to enterprise operating/mission objectives; records established and maintained in accordance with all requirements: SHE incidents, safety tests, etc.
4. Program basis and definition: charter, mission statement, objectives, and strategy; organizational structure, management control and administration, roles and responsibilities, including RASCI (Responsible, Accountable, Support, Consult, Inform), detailed improvement plans and KPIs (Key Performance Indicators), reporting, and methodology for assessing performance and effectiveness.
5. Practices and procedures necessary to implement the Operational Excellence program: risk identification and management, failure, incident and near miss recognition and analysis, and operating value/ROI (Return on Investment) calculation—overall and functional improvement processes.
6. Work culture and institutional values: trust, ownership, support, commitment to excellence, sustainability and continuous improvement, intolerant of defects and inefficiency, organizational transformation and improvement, skills and certification requirements, skills management, training requirements, and documentation.
7. Information management: site facility documentation, specifications, drawings, manuals and records, security and update policy, configuration and revision control, back up, and Management of Change (MOC).
8. Follow up: formal system to measure and monitor performance, periodically and as required; institutionalize and sustain gains, continuous improvement.

Most of the preceding will be applicable to essentially any operating enterprise. As stated, practices and procedures will likely have to be defined for the specific enterprise.

It should be noted that for the purposes of Operational Excellence, asset integrity expands from the conventional definition of pressure containment to assurance of safe operation. Thus, within an operating enterprise, electrical isolation and grounding, insulation and safety devices, turbine overspeed trips, safety and pressure relief valves, steering and brakes on mobile equipment, and crane booms, rigging, and similar equipments are included in the category of asset integrity. It should be noted that most asset integrity professionals agree with this expansion.

Evaluating Results

The assessment template provides a description of best practice for each element and a standardized methodology for evaluating performance and effectiveness compared to best practice across a scoring continuum. This guidance provides a basis for objective assessment and scoring in subjective areas.

Conventional practice has been to establish four maturity levels (quartiles) from no knowledge to best practice. Establishing actual performance and effectiveness begins by ascertaining position within quartiles (one being best) on the basis of a comparison of actual performance compared to descriptions contained in the assessment template. Actual performance is determined by information gained from the preassessment self-evaluation, from interviews and inspection. Identifying first and fourth quartile performance is generally straightforward. If a program element is present and meets most of the stated criteria reasonably well, it is probably in the top, first, quartile. If the operating unit doesn't understand the requirements or understands but have nothing to very little implemented, performance is in the bottom, fourth, quartile.

The middle quartiles: two and three are the most difficult to evaluate and grade. Typically, a program element in the middle two quartiles will meet many or most of the requirements, with some implemented better than others. Because of this, quartiles two and three typically have some but not all requirements met to varying degrees. Assessing performance in the middle quartiles relies on experience and judgment as to how much and how well requirements are met. Experience indicates that for middle quartiles, a well-briefed assessment team will generally agree on a numerical grade within $\pm$ 0.5 on a numerical score of 4.

Necessity for Additional Assessment

The assessment template described so far takes a top level look at elements and results. In some areas, the top level assessment is adequate. In other cases, such as more specialized programs, practices, and procedures, a second or even third level evaluation adding greater detail is necessary for a complete assessment.

In these areas, the top level, evaluated by the Operational Excellence template, principally addresses results and overall context—whether the program is in existence, does it have a written charter, and does it meet overall requirements including scope, application, people with assigned responsibilities, training, and certification. Are there defined reporting and follow-up procedures, and are they followed? Most importantly, is the program effective and achieving results?

The next level examines a program in more detail, including application, set up, organization, management system, administration, and implementation. In some cases, there may be a necessity for a third level of detail to assess technical scope and precision in yet greater detail. Second and third level assessments are best conducted from purpose templates following the principles outlined for the overall assessment template.

CONCLUSION, SUMMARY, RECOMMENDATIONS, AND SITE WRAP UP

At the end of the assessment, the team leader compiles information and presents a preliminary report to the operating unit. It is customary and very good relations to deliver a summary report to operating unit leadership, the assessment sponsor, host, and other interested people (selected interviewees) before departing the site. The preliminary report is to confirm the accuracy of observations, summarize findings, outline preliminary conclusions and recommendations, and invite comments and questions.

The summary may be written and/or a visual PowerPoint presentation. The presentation should begin by expressing appreciation for cooperation and thanking the operating unit host for support, arrangements, and cooperation of the people involved.

Observations and conclusions are summarized to include strengths, areas in which the operating unit is doing well, areas for improvement in rough priority, as well as preliminary recommendations.

The site wrap up presentation can include a preliminary performance evaluation for each area assessed. There shouldn't be any numerical scores expressed or even suggested in the departure summary.

The summary should include an invitation for comments on procedure, coverage, questions, observations, conclusions, and recommendations.

PREPARE FORMAL ASSESSMENT REPORT

After the site visit, the team leader consolidates the numerical grades and meeting notes from which to prepare a formal report of the findings, including recommendations for improvements. The team leader may average numerical grades to arrive at a final grade. If there is a wide divergence in numerical grades or notes, the team leader meets with team members to resolve any differences and to reach consensus agreement for the final report.

The formal assessment report is typically developed in five sections as follows:

- Executive summary
- Introduction and objectives
- Conclusions and recommendations

- Narrative
- Appendices.

Executive Summary: Key Findings and Recommendations

This crucial part of the assessment report is drafted last. The executive summary must be brief, no more than one page; summarize all major observations, conclusions, and recommendations related to current performance and improvement with powerful, confirming justification.

Introduction and Objectives

The introduction and objectives section is normally written first. It is advantageously drafted before visiting the operating unit as part of the initial alignment presentation. Use as the alignment presentation validates the process and objectives, assuring all are in agreement at the outset of the assessment process.

Conclusions and Recommendations

This portion of the assessment report is drafted next to last and leads directly to the executive summary.

Conclusions concisely describe performance and effectiveness as evaluated by the assessment team and include numerical performance scores. A graphical presentation of the scores is advantageous either in the conclusions and recommendations section or in a referenced appendix. Radar plots are recommended, although the exact choice is up to the assessment team leader and should be the graphic presentation most familiar to the operating unit being assessed. Every score must be supported by a narrative explaining the score. Areas and activities performing well are noted.

Opportunities for improvement (gaps) defined as variations from best practices that are identified during the assessment are noted in the conclusions section. Opportunities for improvement, more specifically potential gain by implementing improvements, are stated in the conclusions as a means to prioritize improvement actions. Opportunities for improvement are linked to specific proposed improvement actions in the recommendations section.

It is a good idea to discuss opportunities for improvement with the operating unit during the report drafting process. This to assure alignment and agreement or, as a minimum, issues understood before the final assessment report is issued.

Within the conclusions and recommendations, recommended actions are most important as are the basis for improvements.

Narrative

The narrative is written directly from consolidated interview notes organized in logical categories. As an example, the organization, administration, and control processes may be covered in most interviews. For the purposes of the narrative section, all observations and comments pertaining to organization would be compiled into one section of the narrative.

The narrative constitutes the bulk of the report and must support and justify all conclusions and recommendations. Numerical grades may be in tabular form in the narrative or in an appendix.

Appendices

Appendices contain supporting documentation used in the assessment and report including documentation supplied by the operating unit. Appendices are included as necessary to make the assessment a self-contained, auditable document that accurately establishes conditions at one point in time.

If an assessment template containing questions and best practices is prepared and used, it should be included as an appendix.

It is highly recommended to have the operating unit review the narrative, conclusions, and recommendations before final issue. This will assure accuracy, identify, address, and correct any misunderstandings, misinterpretations, and potential disagreements.

REPORT SUBMISSION

On completion, the assessment report is submitted and/or presented to operating unit leadership and to selected personnel (typically all interviewed and the operating unit site host). All are asked to review the assessment report, including findings, conclusions, and recommended actions. Any questions and recommendations for suggested changes should be addressed to the assessment team leader for consideration for inclusion in a revised report.

The assessment team leader is the final authority for all suggested changes.

ACTIONS REQUIRED FROM ASSESSED OPERATING UNIT

Leadership

Operating unit leadership is responsible for prioritizing recommendations for improvements and appointing and supporting the Operational Excellence program organization described in Chapter 10 to carry out the improvements. If the Operational Excellence program is not yet established, the assessment begins the process. If an Operational Excellence program is operational, the assessment refines and potentially reprioritizes opportunities for improvement and improvement action plans detailed in previous chapters.

FORMAL AND INFORMAL EMPLOYEE SATISFACTION SURVEYS

Before concluding this chapter on assessments, there is another area that must be mentioned: employee satisfaction surveys.

Employee satisfaction surveys provide a means to evaluate attitudes and ensure that the sense of employees and the working culture, good and bad, are identified, understood, and corrected as necessary. Surveys provide a positive vehicle for suggestions and a means to legitimize complaints. Employee surveys are typically developed and implemented as part of the communications program; refer to Chapter 10. They should be performed regularly within the framework of an overall communications strategy and designed to develop specific information.

A single survey should have no more than five or so basic questions centered on a central theme that is being examined. Too many questions and responders will either not answer at all or not in the detail necessary.

Surveys can be developed and conducted in-house or by an experienced third party. The latter may be more effective if participants feel safer and are more open making candid comments to a third party rather than someone in their own organization. Anonymous comments are allowed to encourage candid evaluations and comments. When employees sign their name, they should be contacted to express appreciation for their interest, comment, and willingness to identify themselves and to ascertain if there might be more.

Having completed the survey, it is essential the results are publicized and communicated as soon as practicable. Communications should include an evaluation of responses and conclusions, identifying the positive as well as any identified need for additional improvement. The survey must result in a formal action plan defining tasks and any changes required to address and correct the major issues.

A survey increases awareness among participants and elevates expectations that real changes will be made. The survey must be followed quickly by a detailed summary of results, evaluation, and commitment to corrective action and an action plan if required. If not followed by any discernable action, people become cynical and conclude the survey is simply another exercise where management wants to feel good but doesn't really want to hear or even care about what is actually going on within the organization

For these reasons, it is best not to conduct a survey unless leadership is prepared to listen to and act on the survey's results.

> If you talk to the people who are directly involved—ask them what to do about something—and then do what they said, your operation normally runs better.

WHAT YOU SHOULD TAKE AWAY

The assessment process provides an objective, in-depth examination of an operating site, facility, function, or process to evaluate performance, efficiency, and results at one point in time. Actual performance is compared to operating/industry best practice to gain an array of prioritized opportunities for improvement. The assessment process is designed and organized to fit specific site/facility business/mission objectives, as well as objectives and requirements for specific processes and programs.

Although assessments are often thought of as a means to evaluate the performance of an ongoing improvement initiative or program, they are exceptionally valuable to establish current conditions and identify and prioritize improvement opportunities for a new improvement program.

The assessment is conducted in a disciplined, well-defined sequence to assure all relevant facts and information are identified and investigated in sufficient depth to assure full understanding.

A proven and tested assessment template is highly advisable. A template will assure all relevant areas; applicable processes and activities are covered in sufficient depth. A well-constructed template will have sufficient detail and comparative evaluation criteria to maximize objectivity. Finally, the template assures that all participating in the assessment are working from a common base.

The assessment itself is conducted by an experienced assessment team and team leader.

The assessment will identify strengths, areas in which current performance and results are equivalent to operating/industry standards, as well as opportunities for improvement prioritized by potential gain. Evaluation and scoring criteria must be consistent and supported by observations.

The assessment is documented by a complete report. Observations and recommendations are explained and displayed in a format and terms that will be easily interpretable by operating/site personnel. The report serves as closure to the assessment; an objective, auditable evaluation of performance, efficiency, and results at one point in time, as well as a reference from which to compare subsequent performance.

18

CHECK—MEASURE AND MANAGE RESULTS

The Check and Improve processes described in this chapter and the next are tightly linked and of continuous duration in an Operational Excellence program. They connect and work together to assure maximum results and that the processes and practices responsible for producing results are sustained. Simply stated, Check is a continuing process to measure results for assurance they meet or exceed expectations for contribution to enterprise and program business/mission objectives, value, return, and compliance to schedule.

In practical terms, even the best, most well thought out, detailed, and implemented plan will meet unexpected challenges and barriers. To paraphrase a military axiom: "plans are perfect until first placed in action—then reality introduces itself!" Learning and planning do not stop at doing; rather both continue. The unexpected should be expected. The Check process identifies the unexpected. Improve includes the adjustments, refinements, and activities to assure gains are firmly embedded into the institutional cultural fabric.

> A comment was made during a workshop that the enterprise was quite good at defining actions, including detailed steps to be taken. Often, this did not carry over to implementation. Many cited major weaknesses in follow-up to assure anticipated results are achieved and the improvement is sustained.

Operational Excellence: Journey to Creating Sustainable Value, First Edition. John S. Mitchell.

BEGIN WITH METRICS

As discussed in Chapter 14, each improvement action plan must have several activity and results Key Performance Indicators (KPIs); including time milestones. Properly designed KPIs assure improvement initiatives are proceeding on schedule and producing objective results. KPIs must be capable of quickly identifying deviations from plan, including time.

Results metrics are established for each improvement initiative to demonstrate performance gains in operating, process, and system effectiveness. They must cascade into overall program and enterprise business/mission objectives to demonstrate value gain, preferably financial or another top-level metric that is credible to senior executive and financial management. Compelling assurance that objectives are being met and demonstrable business/mission value gained is essential to attract and maintain high level executive support.

Activity metrics, as explained in Chapter 15, should be employed to assure action plan prerequisites such as reorganization, training, improvement/insertion/deployment of practices, and technology are being completed on schedule.

CONDUCT ASSESSMENTS AND SURVEYS

The Check process begins with periodic evaluation of the performance and results of specific improvement initiatives; refer to Chapter 17. All performance metrics: activities, results, and time to achieve are measured and monitored, at least monthly. As stated earlier, it is essential to assure compliance to expectations. The evaluation centers on effectiveness and quality metrics to judge success and identify areas where adjustments in the improvement plan and/or further training might be necessary. As part of the process, the Action Team must review the original objectives, improvement plan, and organization, to assure their solution is optimal and contributing the greatest amount possible to business/mission effectiveness and increased value.

Timely monitoring assures barriers to success, deviations, and shortfalls in performance are spotted quickly. With barriers, deviations, and shortfalls identified, action plans can be adjusted and improved to overcome unforeseen issues and challenges, with least impact on results or schedule. Wherever possible, team leaders should endeavor to get ahead of plan to provide ample time to correct for unforeseen challenges and/or shortcomings.

> Successes, gained by teams, that have expended great effort to achieve results, must be celebrated in communications to increase the enthusiasm, ownership, and support for the necessity and value of improvement and the Operational Excellence program.

To facilitate understanding and promote ownership, responsibility, and support, displaying performance metrics in a graphic 1-year running format has proven very effective; refer to Figure 18.1.

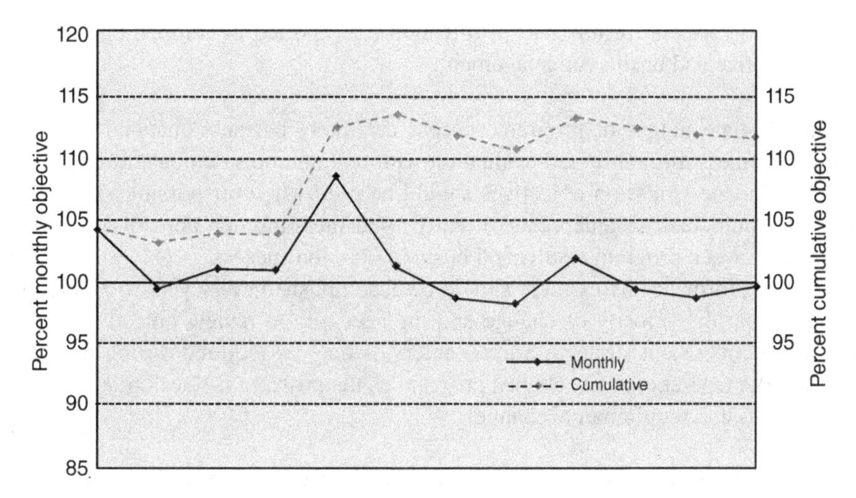

Figure 18.1 Twelve-month rolling and cumulative display of a performance KPI.

As improvement initiatives mature, assessment intervals may be lengthened depending on factors such as pace of improvements, probability of meeting all expectations, and proximity to final, objective results.

Performance reviews often disclose additional high value opportunities for improvement that can be inserted into the Operational Excellence process for consideration.

Formal Performance Assessments

Regular, scheduled performance assessments and employee surveys as detailed in Chapter 17 can be considered part of the Check phase. Summarizing, the objectives of a performance assessment are to review, validate, and/or update program objectives and strategy—overall or specific Operational Excellence plans, performance, and results including as applicable:

- Validate governing enterprise business/mission objectives and conformance of the Operational Excellence program strategy.
- Assure the enterprise vision and objectives—the Operational Excellence program objectives, action, results required, and benefits are fully understood and accepted.
- Confirm the Operational Excellence program aligns with, is effective, and producing demonstrable value and value contribution to enterprise business/mission objectives.
- Assure program, processes, and individual improvement initiatives are effectively producing value and on-time results.
- Identify additional opportunities for improvement.

- Assure financial results and justification are reported in compelling terms to executive and business management.

The ultimate judges of program success are likely business/financial oriented. Thus, a business/mission-based evaluation with real numbers demonstrating contribution to business/mission objectives should be used wherever possible. Likewise, program results metrics must relate directly to business/mission objectives to prove the bond between program results and business/mission success.

A full performance assessment should be accomplished every year to 18 months depending on the velocity of change and the necessity to review objectives, foundation principles, and strategy. Shorter intervals may be required during the early stages of the Operational Excellence program as the program is developing or when business/mission requirements change.

One facility with an improvement program in place conducts informal reviews every 6 months. Formal reviews are conducted annually, with revisions made as required after the annual review.

The review process must recognize that a changing business/operating environment most often elevates requirements for performance and effectiveness. Best, "benchmark" facilities are continually improving performance. What was excellent last year may be average the next year! Continuing pressure on profit margins typically results in corresponding pressure to reduce cost. Often, the Operational Excellence program can conclusively demonstrate greater benefits and value from improved effectiveness.

The review must also provide continuing assurance that resources are being allocated to the right activities at the right time to gain optimal business/mission effectiveness. To strengthen this aspect of Operational Excellence, business/mission objectives must be infused into day-to-day management decisions.

CONFIRM RESULTS AND CONTRIBUTION TO ENTERPRISE VALUE AND STRATEGY

As part of the Check process, it is essential to periodically review the overall Operational Excellence program, including the vision, strategy, objectives, requirements, actions, and results. All must be in full alignment and consistent with enterprise business/mission objectives and strategy. All factors that influence the strategy and successful implementation of improvement plans are validated. Plans are reviewed to assure all elements remain effective and consistent with the strategy and objectives. During the review process, every effort is made to identify and value prioritize additional improvement opportunities.

Progress and results—improvements accomplished and value created, are communicated to action teams and publicized throughout the site by the Communications Team, as mentioned in Chapter 10, to build and reinforce support and consensus.

Misunderstandings uncovered during the follow-up process are corrected and communicated.

CONTINUE CHECKING UNTIL CONFIDENT THAT IMPROVEMENT IS FULLY SUSTAINED

Finally, a highly important caution: many improvement initiatives have failed because the process is considered complete as soon as the action plan is implemented. The team, or at least a core group of the team, must remain together, active and meeting periodically (typically at lengthening intervals as the actions mature), to evaluate performance, conduct ongoing assessments of progress to objectives, and develop any refinements or alterations that may be necessary. The team's work is not complete until the improvement is institutionalized—a process that typically requires years. And this leads to the next chapter.

WHAT YOU SHOULD TAKE AWAY

Activity and results metrics for all improvement initiatives are measured and monitored at least monthly to demonstrate progress to objectives and assure results are on target to achieve business/mission/program objectives.

In addition to monitoring improvement initiatives, formal, scheduled performance assessments are conducted every year to 18 months to validate alignment of the program strategy and results to business/mission objectives, confirm the program is producing a demonstrable improvement in value, and identify additional opportunities for improvement. It is imperative that scheduled assessments are performed in accordance with a documented scorecard to assure consistency and accurate evaluation of performance.

It should go without saying that success requires benefits for all involved. The question "What's In It For Me?" must be answerable in the affirmative at every level in the organization for continuing success.

19

IMPROVE—INSTITUTIONALIZE AND SUSTAIN GAINS

The Improve process follows Check and is of continuous duration in an Operational Excellence program. The Improve phase assures results and the processes, practices, and technology that produced the results are continuously improved and sustained. Improve is a continuous process to measure performance and institutionalize improvements that contribute to enterprise and program business/mission objectives. Improve demands a mindset that the status quo is never sufficient, and continuous improvement must be the norm. The Improve process ratchets to assure gains are sustained.

The Check process identifies the unexpected. Improve includes adjustments, refinements, and activities to firmly embed gains into the institutional cultural fabric and feedback information into the Identify phase of Operational Excellence for continuous improvement.

IMPROVE AND SUSTAIN

Improve and Sustain is the second of two continuous processes in an Operational Excellence program. It requires a mindset in the work culture that maintaining gains while continuing to improve is essential; status quo and restoration are unacceptable.

The Improve and Sustain phase of the Operational Excellence program continues from the Check stage and is not unlike the Plan stage. Results are reviewed and activities adjusted as necessary to maintain momentum, improve results, sustain gains, and assure all elements that have produced success are institutionalized.

Operational Excellence: Journey to Creating Sustainable Value, First Edition. John S. Mitchell.

As improvements are achieved, it is crucial to maintain visibility, focus, and pressure for continuing results. As mentioned in the previous chapter, relaxation and early demobilization before gains have been fully realized and institutionalized within the work culture as common failures of typical improvement programs. Sustaining and institutionalizing gains are primary objectives of Operational Excellence.

CONTINUOUS IMPROVEMENT

The Improve phase of Improve and Sustain identifies additional opportunities for improvement that move up in the priority sequence as action plans are fulfilled and gain objective results. Improvement opportunities discovered during the Improve and Sustain phase are inserted into the Identify phase of the program.

Assumptions are continually reviewed and tested. Has the environment changed; have market, business, and/or operating conditions changed? Have objectives changed? Results are also continually monitored to assure compliance with industry-best benchmark performance. If benchmarks have improved/are improving, can effectiveness be improved incrementally by alterations to an existing action plan, or must there be another improvement initiative altogether to meet the new conditions? If so, the appropriate opportunity is formulated and added to the Identify process, completing the circular implementation process. Improvement is continuous; measured results, mission compliance, and profitability improve. This is the process of continuous improvement.

Continue Follow-Up: Adjust/Refine, Improve, and Extend Improvement Action Plans

Periodic and specific monitoring of results, initiated in the Check stage, continues throughout the Improve and Sustain stage. The idea is to build a work culture committed to improvement, continuously evaluating progress to assure results equal or, preferably, exceed expectations. Improvement plans are adjusted, refined, and extended as required to produce results as quickly and efficiently as possible. As part of this process, any additional activities, tasks, and actions beyond those identified in the sustainability plan are identified to maintain, sustain, and build on improvements.

As stated earlier, the follow-up process isn't limited to planned improvements. During the Improve and Sustain process, it is imperative to continue to identify deficiencies, weak areas, and further opportunities for improvement. As additional opportunities are identified, they are inserted into the Identify process for review, refinement, and value prioritization.

Throughout the process, it is also imperative to assure the chain of responsibility for results and time remains unbroken. Responsibility by name and time is assigned for all actions to adjust, refine, improve, and extend improvement action plans.

Expand and Increase Ownership, Responsibility, Accountability, and Commitment

The Operational Excellence program requires full commitment and ownership for improvement and achieving success—nothing succeeds like success! Demonstrating real improvements and progress toward an important objective creates enthusiasm, ownership, and identification of additional opportunities for improvement. In many cases, people observing the enthusiasm and success developing in a pilot improvement project will ask: "When can I/we participate?" When this happens, the facility is on the road to real, sustainable success!

Continue Training

As stated in Chapter 10, training is a vital and continuing part of the Operational Excellence program. The necessity and content for training emerge throughout the program. Specific training requirements are defined by results and the skills and knowledge necessary to achieve results. Training must be directed to spreading knowledge; assuring competency to gain objective results is resident in processes, systems, and procedures and not dependent on individuals.

Training includes coaching for teams to assure they operate effectively and at maximum performance. Team building, strengthening working relationships, and stressing necessities for vital activities such as idea generation, planning, and follow-up are essential.

ACHIEVING SUSTAINABILITY

The comprehensive Sustaining (Control) plan described in Chapter 14 must be constructed into the improvement process at the Plan stage. The Sustaining plan assures that improvements are stabilized and institutionalized and momentum and gains achieved by the Operational Excellence program are continued and improved.

Organizational attributes necessary for sustainable success include a stable, effective, disciplined, and focused work culture. Everyone is strategy oriented, constantly seeking improvements, particularly those capable of delivering quick results within a long-term strategy for success. "Programs of the Month" that create confusion and cynicism are banished! Decisions are delegated consistent with risk and program objectives to assure everyone has "skin in the game." Fair, honest, and open communications about the program, objectives, and even problems and shortcomings are essential.

As results are gained and necessary actions decline, the team responsible for a successful improvement can be reduced in numbers. It is most important for a key member or members of improvement action teams to remain actively engaged in an oversight role, to maintain supervision and responsibility for the systems, processes, and practices that produced the improvements. This will assure results do not regress until improvements are sustained and fully institutionalized.

One major pitfall, temptation, must be resisted: prematurely declaring victory, terminating efforts, and demobilizing the improvement team before results; the systems, processes, and procedures that produced results, are fully institutionalized. As an improvement initiative approaches completion, demobilization occurs in stages to assure gains are not lost. Full demobilization and termination of efforts must not occur until totally confident that improvements are sustainably institutionalized.

The Sustaining plan is reviewed periodically to assure the plan is being followed and is effective. As in other reviews, modifications and refinements are made as necessary to assure results, successes, and the processes and procedures responsible are permanently embedded in the work culture and followed. The review includes assigned responsibility, performance to KPIs (Key Performance Indicators), and continuing training—all to assure improvements are institutionalized and sustained.

Institutionalize Success

The final step in the Operational Excellence improvement process is to institutionalize the concept, process, and successes.

Processes, programs, and procedures are sustained and institutionalized when the new, more effective processes and practices are "the way things are done." There is no memory remaining of the "old" ineffective processes. Management, champions, and team members can be promoted for outstanding results without any loss of momentum or diminishing results.

Full institutionalization, closure with improvements solidly imbedded within the work culture and sustaining, may require 5 years or even more; certainly several times greater than the time required for an improvement initiative. With this stated, the facility will be gaining tangible rewards from program inception!

> Real, sustainable improvement requires permanently embedding the most effective practices and technology into the entire organization!

Communicate and Publicize Progress and Results

Communications are and remain an essential part of and significant contributor to the success of an Operational Excellence initiative. As the excitement resulting from success fades into distant memory, there is a need to keep the people involved enthusiastically committed; continually seeking improved performance and effectiveness. The communications program described in Chapter 10 must be kept alive and evergreen. Publicize results with individual credit to maintain enthusiasm and commitment.

The Communications Team is an essential part of this effort, celebrating and promoting successes and benefits and identifying suggestions and potential areas for additional improvement to create positive visibility, gain support, and ownership.

It is imperative that everyone thoroughly understands the personal and organizational necessity and benefits of increased efficiency and performance. Since success produces success, it is important to publicize improvements accomplished and value created in terms that will create enthusiasm and motivation for the program. This includes identifying and publicizing team and individual successes to create visibility, elevate enthusiasm, and increase support and momentum. The concept, necessity, and benefits of a continuous improvement work culture are reinforced. The Communications Team must be sensitive to and alert for concerns and misunderstandings. Both must be addressed and corrected immediately.

A program leader was attempting to congratulate people by name for contribution and tasks well done in a bi-weekly newsletter. All seemingly went well until individuals began counting words. Several began complaining if congratulations to another individual exceeded the word count of their latest congratulations!

Overcome Resistance

Regardless of how well an Operational Excellence program is formulated; the necessity and benefits communicated, there will always be some resistance. When any changes to current procedure and practice are introduced; no matter how necessary or beneficial, a status quo mentality is often encountered. The status quo mentality can be summarized: "I've been performing this task this way for 20 years, never before until now informed that performance was inadequate, improvements were necessary so I'll just continue in the old way until this program blows over like all the previous have."

Again, communications are essential. The Communications Team and all others engaged in the program must be sensitive to the status quo mentality and take active steps to transform concerns to support. In addition to addressing specific concerns, communications must emphasize that changes in external conditions—business, mission requirements, competitive and regulatory environment, and other factors, necessitate improvements. This includes the necessity to perform tasks more effectively and understand and live by new roles and responsibilities. To assure success, continuing employment, the new, optimum way must be accepted as the only way!

It must be emphasized that compensation, rewards, and peer pressure are high on the list to assure support and success. When introduced to any improvement initiative, most, if not all, employees will say "terrific but what's in it for me (WIIFM)?" WIIFM must be addressed and answered from the outset for every employee; especially if the improvements will ultimately result in fewer employees. Some benefits to be considered and emphasize include the following:

- Safer workplace and greater job security
- Better relationships within the operating unit; greater teamwork and fewer adversarial conflicts
- Flattened organization and increased opportunities to use initiative and take responsibility (it is important to note that some may not want increased responsibility; it is imperative to make them as comfortable in the new organization as people who will thrive in the new conditions)
- Increased compensation based on successful improvements in effectiveness
- Functional silos discouraged; communications, mutual support, and teamwork encouraged.

During a discussion to summarize requirements and benefits of Operational Excellence, one of the participants concluded that the idea of production leading Operational Excellence improvement action teams implied that Production Operations would, for all practical purposes, take over maintenance; an action with which he totally disagreed. He was reassured that a production manager might well be in charge of the Program Leadership Team. However, their task was primarily to identify, value prioritize, and develop actions for improvement. Maintenance remained an independently managed function, and retained its identity and organizational structure with responsibilities for contributing to the improvement action team process and implementing improvement actions as appropriate. Joint participation in action teams did not in any way mean that production was taking over maintenance. They did establish and reinforce the essential notion of an operations/maintenance partnership that is essential to gain maximum success. He expressed satisfaction with the answer and concept.

Results-Based Compensation

Results-based compensation that rewards compliance and results within the new scheme can be a key factor in institutionalization. In some cases, it is absolutely essential. One example of the latter is overtime. If a significant number of employees depend on overtime as part of their compensation, one can imagine the enthusiasm and support that will be received by an improvement initiative to improve effectiveness and reduce overtime! There is a way around this and similar dilemmas. Reward employees with a monetary bonus for reducing overtime. If the bonus is sufficient, equal to approximately half of the lost overtime, everyone wins. The enterprise from improved effectiveness and reduced overtime costs; employees by minimally affected compensation and increased family time.

SOME LESSONS LEARNED

Following are some lessons learned by enterprises that have instituted major improvement initiatives:

- Assure leadership is committed, aligned, and on the same page
- Advantageous for the manufacturing/production function to own the improvement process
- Gain acceptance across the entire organization, including supporting functions: engineering, finance, HR, and IT
- Assign the best people to the improvement program
- Formulate and implement a solid process of improvement
- Assure management stability during the early stages of the program and until results and successes have been demonstrated
- Implement the program and included processes for greatest effectiveness and results
- Maintain focus on value to business/mission objectives
- Communicate frequently at all levels
- Listen to and consider feedback; don't assume all is well
- Hard choices will have to be made

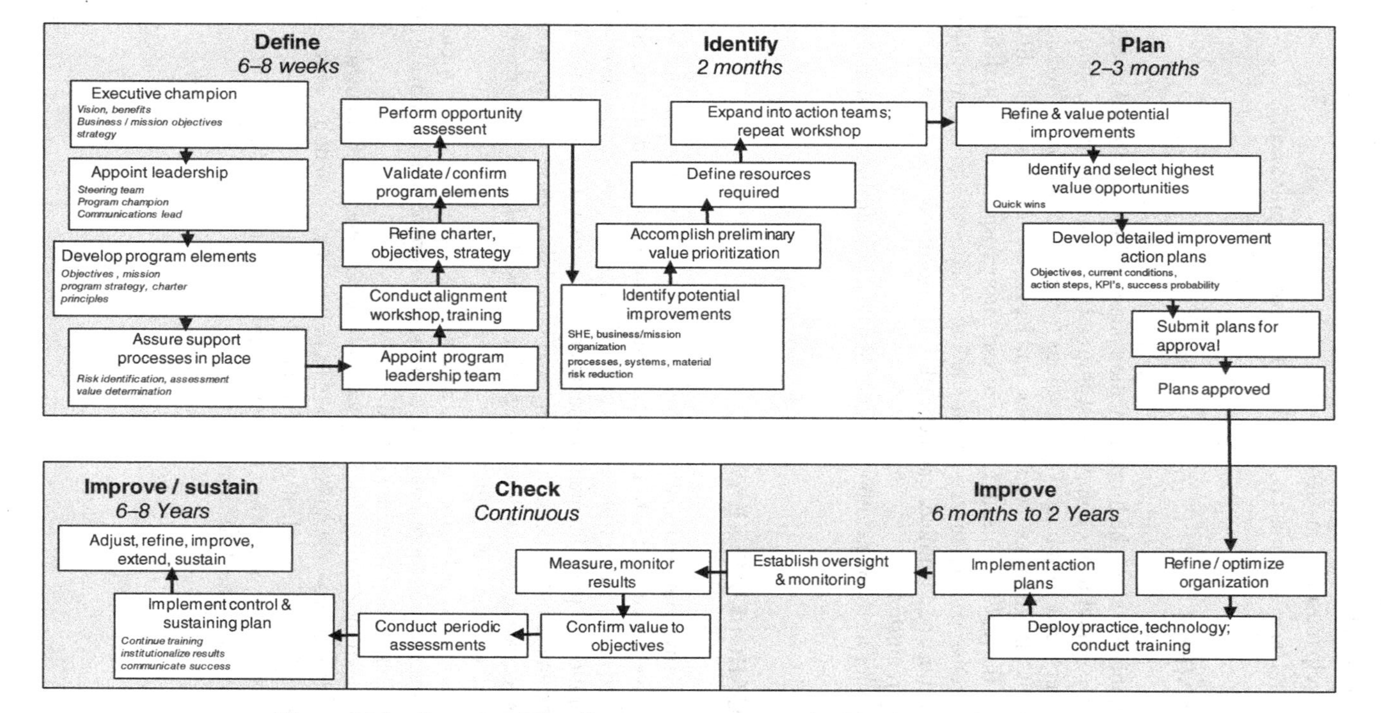
Define
6–8 weeks
Executive champion
Vision, benefits
Business / mission objectives
strategy
Appoint leadership
Steering team
Program champion
Communications lead
Develop program elements
Objectives , mission
program strategy, charter
principles
Assure support
processes in place
Risk identification, assessment
value determination
Perform opportunity
assessent
Validate / confirm
program elements
Refine charter,
objectives, strategy
Conduct alignment
workshop, training
Appoint program
leadership team
Identify
2 months
Expand into action teams;
repeat workshop
Define resources
required
Accomplish preliminary
value prioritization
Identify potential
improvements
SHE, business/mission
organization
processes, systems, material
risk reduction
Plan
2–3 months
Refine & value potential
improvements
Identify and select highest
value opportunities
Quick wins
Develop detailed improvement
action plans
Objectives, current conditions,
action steps, KPI's, success probability
Submit plans for
approval
Plans approved
Improve / sustain
6–8 Years
Adjust, refine, improve,
extend, sustain
Implement control &
sustaining plan
Continue training
institutionalize results
communicate success
Check
Continuous
Measure, monitor
results
Confirm value to
objectives
Conduct periodic
assessments
Improve
6 months to 2 Years
Establish oversight
& monitoring
Implement action
plans
Refine / optimize
organization
Deploy practice, technology;
conduct training

Figure 19.1 Operational Excellence program summarized implementation time line.

- Improvement will take time
- Success is a team effort
- Outside help must be considered
- Results and successes must be heavily publicized and promoted.

FINAL COMMENT

As a final comment to close the Implementation phases of an Operational Excellence program, it is essential to keep uppermost in mind that every operating unit, every facility, and every area are unique. Opportunities for improvement are different, people are different, and strengths, weaknesses, and barriers to success are all different. Only you have the real answers, the ability to sort through opportunities to define the best, most effective path to success. Your specific conditions and opportunities for improvement must be identified, enumerated, and prioritized to achieve success in your special circumstances.

A time line for a complete implementation through sustainability is illustrated in Figure 19.1.

WHAT YOU SHOULD TAKE AWAY

Continuous improvement and sustainability don't just happen. They require a plan, continuing surveillance, and effort to assure gains are maintained, improved, and institutionalized. Continuing follow-up is essential to assure any regression or shortcomings identified by periodic measurements and assessments are corrected. Every effort must be made not to declare victory and demobilize an improvement initiative before processes and procedures that produced results are fully and irreversibly institutionalized into the work culture.

Communications to publicize success and benefits continue as does training to assure processes and procedures introduced during an improvement initiative are reinforced, strengthened, and institutionalized.

It should go without saying that full success requires benefits for all involved. To assure continuing success, the question "what's in it for me" must be answered in the affirmative, satisfactorily to all, at every level in the organization.

Total success comes only when the new, improved way becomes the only way; there is no memory of the "old" way.

20

CONCLUSION—NOW IT IS UP TO YOU!

So there you have it: a roadmap to successful Operational Excellence that is adaptable for every enterprise and every journey. Is everything expressed in the previous 19 chapters totally applicable to you? Can all the challenges you may face in your journey be predicted? Probably not to both questions. But use the concepts, principles, and ideas as a starting point—the guidance necessary to give you confidence of ultimate success.

You'll undoubtedly start out simply attempting to remain on the road to improvement; staying out of the ditches. There will be unexpected surprises and problems; storms and challenges. Establish the right leadership and build a positive committed ownership and success-oriented work culture, all obstacles will be surmounted. As you progress and improvements take hold, you'll undoubtedly find the journey getting easier. People will be happier and contribute more. Instead of simply attempting to stay out of ditches, you will realize the road is solid and your vehicle is capable and steering optimally. Gaining greatest value and return is easier. Operational Excellence and the Operational Excellence program provide the confidence and forward visibility to assure your success.

Very best wishes to achieve the great success that is within your grasp!

Operational Excellence: Journey to Creating Sustainable Value, First Edition. John S. Mitchell.

INDEX

Operational Excellence: Journey to Creating Sustainable Value, First Edition. John S. Mitchell.